SYSTEMS ENGINEERING FOR COMMERCIAL AIRCRAFT

For as the body is one, and hath many members, and all the members of the body, being many, are one body ... For the body is not one member, but many. ... But now are they many members, yet but one body. And the eye cannot say unto the hand, I have no need of thee, nor again the head to the feet, I have no need of you. ... And whether one member suffer, all the members suffer with it; or one member be honoured, all the members rejoice with it. [I Corinthians 12: 12–26, Holy Bible, *King James (Authorized) Version]*[1]

---

1   Note: The above quotation is included for its applicability to systems theory and not for any theological reason.

Systems Engineering for Commercial Aircraft

A Domain-Specific Adaptation

Second Edition

SCOTT JACKSON

*University of Southern California and
Burnham Systems Consulting, USA*

Routledge
Taylor & Francis Group

LONDON AND NEW YORK

First published 2015 by Ashgate Publishing

2 Park Square, Milton Park, Abingdon, Oxon OX14 4RN
711 Third Avenue, New York, NY 10017, USA

Routledge is an imprint of the Taylor & Francis Group, an informa business

First issued in paperback 2017

British Library Cataloguing in Publication Data
A catalogue record for this book is available from the British Library

The Library of Congress has cataloged the printed edition as follows:
Jackson, Scott, author.
  Systems engineering for commercial aircraft : a domain-specific adaptation / by Scott Jackson.
    pages cm
  Second, revised edition of: 1997.
  Summary: "Explains the principles of systems engineering in simple, understandable terms and describes to engineers and managers how these principles would be applied to the development of commercial aircraft"--Provided by publisher.
  Includes bibliographical references and index.
  ISBN 978-1-4724-3921-5 (hardback)
1. Aeronautics--Systems engineering. 2. Transport planes--Design and construction.
I. Title.
  TL671.2.J18 2015
  629.135--dc23

2014039548

ISBN 978-1-4724-3921-5 (hbk)
ISBN 978-1-138-04529-3 (pbk)

Contents

List of Figures

List of Tables

Acknowledgments

For the first edition, I continue to be grateful to Jim Kehres who, in about 1963, advised me "to learn more about systems engineering." I am also especially grateful to Gary Burgess who, at the Douglas Aircraft division of McDonnell Douglas, saw the need for systems engineering in commercial aircraft. The following people were especially helpful in the preparation of this book: First, Tom Nagle and Archie Vickers were helpful for their overall knowledge of the systems engineering process. In addition, the following people contributed much: Gary Bartz, certification; Peter Camacho, top-level sizing of aircraft; Darlene Carpenter, electrical; Madrona Geisert, interior systems; Fred Gray, propulsion; Stu Hann, safety and reliability; Don Hanson, aircraft development; Brian Keeley, airframe; Dr Noreen McQuinn, human factors; Christine Ostrowski, maintainability; Mo Piper, avionics and software; Bob Rich, functional analysis; Todd Strong, mechanical systems; Matt Vance, QFD; Steve Wiles, environmental control; and Beth Clark, integration and life-cycle analysis. Finally, my wife Carole provided invaluable advice regarding syntax, diction, and organization.

For the second edition I am indebted to Jim Hines for further insight into functional analysis; John Hart-Smith, composite structures; Ashok Jain for the supplier perspective; Derek Hitchins for systems theory and cluster analysis; Ed Conrow for risk management; Tim Ferris for assistance with resilience analysis; Wellington Oliveira for the regional jet perspective; and Karin Mayer for help with grammar, spelling, and diction.

Acronyms and Abbreviations

Term	Definition
AC	Advisory circular
AC	Alternating current
ACARS	ARINC communication addressing reporting system
ADI	Attitude direction indicator
AFM	Airplane flight manual
AIT	Analysis and integration team
ALAR	Approach and landing accident reduction
APU	Auxiliary power unit
ARINC	Aeronautical Radio Incorporated
ARP	Aerospace recommended practice (SAE)
ASA	Airplane state awareness
ASAP	Aviation safety action plan
ATA	Air Transport Association of America
ATC	Air traffic control
AVOID	Airborne volcanic object identifier and detector
BCD	Baseline concept document
BFE	Buyer furnished equipment
BITE	Built-in test equipment
BWB	Blended wing-body
CAST	Commercial Aviation Safety Team
CCA	Common cause analysis
CDE	Chief design engineer
CDR	Critical design review
CEO	Chief executive officer
CFD	Computational fluid dynamics
CFIT	Controlled flight into terrain
CM	Configuration management
CMR	Certification maintenance requirement
CONOPS	Concept of operations
COTS	Commercial off-the-shelf products

Term	Definition
c.p.	Center of pressure
CRM	Cockpit (or crew) resource management
CSD	Central speed drive
CSE	Chief systems engineer
DC	Direct current
DCAS	Digital core avionics system
DF	Development fixture
DFMA	Design for manufacture and assembly
DME	Distance measurement equipment
DOC	Direct operating cost
DOORS	Dynamic Object-Oriented Requirement System
DOS	Director of safety
ECS	Environmental control system (or subsystem)
EDF	Electronic development fixture
ELB	Emergency locator beacon
ELT	Emergency locator transmitter
EMI	Electro-magnetic interference
ESE	Enterprise systems engineering
ESR	Engineering safety review
EWO	Engineering work order
FAA	Federal Aviation Administration
FAR	Federal aviation regulation
FBL	Fly-by-light
FBW	Fly-by-wire
FCA	Functional configuration audit
FDA	Food and Drug Administration
FEA	Finite element analysis
FFBD	Functional flow block diagram
FFRR	First flight readiness review
FFR	First flight review
FHA	Functional hazard assessment
FMS	Flight management system
FOD	Foreign object debris
FOQA	Flight operational quality assurance
FTA	Fault tree analysis
G&A	General and administrative [costs]

Term	Definition
GMT	Greenwich mean time
GPS	Global positioning system
GPWS	Ground proximity warning system
HBPR	High by-pass ratio
HF	High frequency
HIRF	High-intensity radiation field
HSCT	High-speed civil transport
HSI	Horizontal situation indicator
HUD	Heads-up display
ICA	Initial cruise altitude
ICD	Interface control drawing (or document)
IDEF0	Integrated definition for function modeling—type 0
IDG	Integrated drive generator
IEEE	Institute of Electrical and Electronic Engineering
IFR	Instrument flight rules
ILS	Instrument landing system
IMACH	Improved methods for aircraft cargo handling
INCOSE	International Council on Systems Engineering
INS	Inertial navigation system
IPD	Integrated product development
IPT	Integrated product team
IVATF	International Volcanic Ash Task Force
JAA	Joint Aviation Authorities
JAR	Joint airworthiness requirements
LRU	Line replaceable unit
LSS	Large-scale system
LSSI	Large-scale system integration
MAP	Maximum allowable probability
MCBF	Mean cycles between failures
MCBUR	Mean cycles between unscheduled removals
MEL	Minimum equipment list
MEW	Manufacturer's empty weight
MMEL	Master minimum equipment list
MMH/1000FH	Maintenance man-hours per 1000 flight hours
MN$/1000FH	Maintenance cost per 1000 flight hours
MSAW	Minimum safe altitude warning

Term	Definition
MT$/1000FH	Material cost per 1000 flight hours
MTBF	Mean time between failures
MTBUR	Mean time between unscheduled removals
MTOW	Maximum take-off weight
MTTR	Mean time to repair
MVA	Minimum vectoring altitude
NDI	Non-development item
NEA	Nitrogen enriched air
NOX	Nitrous oxide
NTSB	National Transportation Safety Board
OEM	Original equipment manufacturer
OSHA	Occupational Safety and Health Administration
PBW	Power-by-wire
PCA	Parametric cost analysis
PCA	Physical configuration audit
PCA	Propulsion controlled aircraft
PDR	Preliminary design review
PRA	Probabilistic risk analysis
PSAC	Plan for software aspects of certification
PSE	Product systems engineering
PSSA	Preliminary system safety assessment
QFD	Quality function deployment
RAS	Requirements allocation sheet
RATs	Ram air turbine
RI	Runway incursion
RNAV	Area navigation
RNP	Required navigation procedures
RTCA	Radio Technical Commission for Aeronautics (former name of RTCA, Inc.)
SAE	Society of Automotive Engineers
SATCOM	Satellite communications
SCM	Software configuration management
SCS	Software configuration index
SDR	System design review
SE	Systems engineering
SEBoK	Systems Engineering Body of Knowledge

Term	Definition
SEIT	Systems engineering and integration team
SELCAL	Selective calling
SEMP	SE management plan
SFC	Specific fuel consumption
SOP	Standard operating procedure
SoS	System of systems
SOW	Statement of work
SSE	Service systems engineering
SQA	Software quality assurance
SRR	System requirements review
SSA	System safety assessment
SVR	System verification review
TAWS	Terrain avoidance warning system
TBD	To be determined (for a requirement)
TCAS	Traffic collision avoidance system
TEAM	Technology evaluation and adaptation methodology
TPM	Technical performance measure
TQM	Total quality management
TRL	Technology readiness levels
UER	Unscheduled engine removals
VFR	Visual flight rules
VGSI	Visual glide slope indicator
VHF	Very high frequency
VMC	Visual meteorological conditions
VOR	VHF omni-directional radio
VSCF	Variable-speed constant frequency

Symbols

AR	Aspect ratio
C_D	Drag coefficient
C_{D0}	Lift-independent drag coefficient
C_L	Lift coefficient
$CL^2/\prod ARe$	Induced drag
C_{LIC}	Initial cruise lift coefficient
C_{Lmax}	Maximum lift coefficient
C_{Lto}	Take-off lift coefficient
ΔC_{Dc}	Compressibility drag coefficient
e	Lift efficiency
Fb	Block fuel
L/D	Lift to drag ratio
$\Lambda_{c/4}$	Sweepback angle at quarter chord
M_{div}	Divergence Mach number
R	Total range
R_{cl}	Climb range
R_{cr}	Cruise range
SFC	Specific fuel consumption
$(t/c)_{ave}$	Average thickness to chord ratio
U	Aircraft utilization factor
V_b	Block speed
$(W/S)_{IC}$	Initial cruise wing loading
$(W/S)_{to}$	Take-off wing loading
W_0	Initial cruise weight
W_1	$W_0 - W_f$
W_f	Weight of fuel
W_{to}	Weight at take-off
W_{wing}	Wing weight
$W_{fuselage}$	The fuselage weight
$W_{landing\ gear}$	The landing gear weight
$W_{nacelle\ \&\ pylon}$	The nacelle and pylon weight
W/T	Thrust loading

W_{TS} The tail section weight

W_{fuel} The fuel weight

$W_{payload}$ The payload weight

$W_{fixed\ equipment}$ The fixed equipment weight

Preface

There have been many developments in the commercial aircraft domain since the publication of the first edition in 1997. From the technology point of view, there have been many innovations, such as the introduction of composite materials and flight envelope protection, both discussed in Chapter 2, among other developments. With respect to safety, the emergence of the Commercial Aircraft Safety Team (CAST), an international consortium of manufacturers, regulators, employee groups, and airlines has served both to track developments in safety and also to suggest improvements in procedures which have reduced the fatality rate dramatically. From a management point of view, the increased use of outsourcing discussed in Chapter 14, has created a challenge for which greater rigor in supplier management is required. This chapter discusses outsourcing in the context of large-scale system integration (LSSI), an advanced topic in the systems engineering lexicon.

In addition to developments in the commercial aircraft domain since 1997, systems engineering has continued to grow in scope and maturity both as a general concept and also in the commercial aircraft domain. The publication of the *Systems Engineering Body of Knowledge* (SEBoK) edited by Pyster (2012) has expanded the scope of systems engineering into three categories: product systems engineering (PSE), enterprise systems engineering (ESE), and service systems engineering (SSE). The discussion of outsourcing in Chapter 14 falls more into the ESE category. Within the commercial aviation domain two important documents have been published: First is the Federal Aviation Administration (FAA) *Systems Engineering Manual* (2014). Secondly, the Society of Automotive Engineers (SAE) guideline ARP 4754A (2010) lays out in a concise way how systems engineering applies to aircraft development with a focus on safety and certification.

This book is not intended to replace the above standards and guidelines or to be a definitive interpretation of them. Rather it is the intent to be a "pointer" to these documents, to show how they can fit into a systems engineering context, and to adapt to these processes as discussed below. Furthermore, this book is not intended to be a manual or handbook; rather it is intended to be a guide to understanding.

If there is a central theme of this book it is that the commercial aircraft domain requires attention to the adaptation of the systems engineering process to that domain. Chapter 13 is devoted entirely to the challenges of adaptation. You may have noticed that this edition is subtitled *A Domain-Specific Adaptation*. These challenges result from the unique demands of the market and the technologies in that domain. In addition, an important fact is that there is already considerable systems engineering in this domain, and the developer can take advantage of that fact by incorporating only those aspects that do not already exist. That chapter also describes how an *existing* organization can perform systems engineering to maximum advantage.

Another goal of this edition is to persuade the developer that systems engineering is not the burdensome process it is often perceived to be in other domains, but rather a logical approach to system development.

An important issue in modern commercial aircraft development is the existence of risks. Although the first edition devoted a subsection to this subject, this edition expands that discussion to an entire chapter, Chapter 15, to the principles of risk management and typical risks that the developer may encounter. With respect to risks, this book does not mention specific aircraft developers, specific aircraft, or specific incidents except to the extent that they are mentioned in accident reports by, for example, the National Transportation Safety Board (NTSB).

A final topic not discussed extensively in other texts with respect to commercial aircraft is *resilience* in Chapter 16; an exception is Hollnagel et al. (2011). Resilience is different from safety in that while safety is concerned with the prevention of failures, resilience deals with the anticipation, withstanding, and recovery from any kind of adverse disruption.

It is hoped that you will find this edition both useful as well as informative regarding the commercial aircraft domain in the context of systems theory and in particular systems engineering.

Scott Jackson
Irvine, California

1

Introduction

The primary purpose of the book is to provide the reader with the information to apply the systems engineering process to the design of new aircraft, derivative aircraft, and change-based designs. A second purpose is to provide guidance that will allow the reader to adapt this process to the commercial aircraft domain through judicious selection of those aspects that would provide the highest leverage of benefits and the lowest risk of adversities. It is assumed that the reader either already has a basic understanding of the process or can obtain that information from further reading of other sources, such as the ones discussed later in Section 1.3. Although there are many interpretations of systems engineering, the principles discussed are generally universal. This book attempts to stress those which are most relevant to aircraft design.

For brevity, the initials SE will be used for systems engineering throughout this book.

1.1 Definition of a System

A system is anything with many parts, like an airplane, a wrist watch, the human body, or the US government. The parts of a system are hierarchical: that is, the airplane parts can be subdivided into subsystems, sub-subsystems, and so forth. However, the principles described here apply equally well to the design of a subsystem as to a system. The official definition of a system adopted by the International Council on Systems Engineering (INCOSE) is as follows: A system is:

> … an integrated set of elements, subsystems, or assemblies that accomplish a defined objective. These elements include products (hardware, software, and firmware), processes, people, information, techniques, facilities, services, and other support elements. (2010, p. 5)

In the commercial aircraft industry the term *system* is normally used for electrical systems, hydraulic systems, and so forth. The term can also refer to the global aviation system. However, in this book *system* will be used in the SE context: that is, for the entire aircraft and its supporting elements. Subordinate elements will be referred to as subsystems, such as the electrical subsystem.

This definition, though, includes non-technical aspects, such as people. It may seem contrary to the classical definition of engineering to include these aspects.

However, this definition is consistent with the modern definition of SE which deals with the effort to define such systems.

1.2 Definition of Systems Engineering

It is difficult for two systems engineers to agree on a definition of SE. There are many definitions and many theories on the implementation of SE. We will look at only a few definitions here.

As the discipline began to take form, the search for a definition also began. The need for such a discipline has resulted from a worldwide trend of devoting an increasingly larger portion of the engineering effort towards pre-design requirements definition. SE is a key methodology in that trend.

The Systems Engineering Body of Knowledge (SEBoK) defines three types of SE: product SE (PSE), enterprise SE (ESE), and service SE (SSE). The SE of an aircraft is product SE. Product in this context can broadly be interpreted to include the operators and maintainers. Product SE is the primary, but not exclusive, focus of this book. Enterprise SE includes the developer, the suppliers, and the carriers. This book does incorporate aspects of enterprise SE, for example, in Chapter 14 which discusses large-scale system integration (LSSI) with an emphasis on supplier management, an important element in enterprise SE.

SE is a discipline which has the goal of arranging the parts of a system in such a way that the entire system does something optimally, such as to get from A to B in a minimum time, or at a minimum cost. That is, SE optimizes the system's performance. However, SE goes even further. Major goals of SE are: first, to define the system's requirements so well that the product will never have to be redesigned; secondly, to make the product as reliable as possible; and finally, to make the customer happy. In practice, these goals may seem impractical. However, SE provides some methods which may bring the design closer to the goals.

The official definition of SE adopted by INCOSE is as follows:

> Systems Engineering (SE) is an interdisciplinary approach and means to enable the realization of successful systems. It focuses on defining customer needs and required functionality early in the development cycle, documenting requirements, and then proceeding with design synthesis and system validation while considering the complete problem: operations, cost and schedule, performance, training and support, test, manufacturing, and disposal. SE considers both the business and the technical needs of all customers with the goal of providing a quality product that meets the user needs. (INCOSE 2010)

This definition raises some important points: First, it points out that SE addresses the entire life-cycle of the system, not just the operational phase. This book addresses the life-cycle functions of an aircraft system in Section 3.1. Secondly, it states that SE assigns requirements to people and processes, not just the aircraft.

Section 5.5 discusses the ability and limitations of assigning requirements to people. Section 5.14 discusses how SE can assign requirements to the fabrication and assembly processes.

An earlier definition was given by Simon Ramo (1973) as follows:

> Systems engineering is a branch of engineering that concentrates on the design and application of the whole as distinct from the parts … looking at a problem in its entirety, taking into account all the facets and all the variables and relating the social to the technical aspects.

Note Ramo's inclusion of social factors in the design of a system. This definition indicates the possible breadth of SE.

The term *systems engineering* has also been used to pertain to computer systems, for example, or to subsystems, such as electrical or hydraulic systems. It will be used in this book in the broadest possible sense, that is, to pertain to any system.

The term *system engineering* has also been used. However, SE has become the industry standard. Although treatises have been written breaking down SE into many steps, let's just consider five, for simplicity, as shown in Figure 1.1. Notice the iterative nature of the first three steps. We will discuss each of these throughout the course of this book. The *FAA Systems Engineering Manual* (2014, p. 5) contains a more expanded view of the SE process.

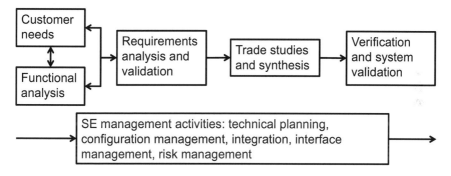

Figure 1.1   Steps in the SE process

Another graphic often used in the description of SE is the famous Vee model. This model is more appropriate when applied to the discussion of requirements; hence, Chapter 4 discusses this model and its implications.

1.3   Historical Background

Early authors to recognize the value of SE and describe the process include A. D. Hall in his book *A Methodology for Systems Engineering* (1962). Other early descriptions of the process include the Army Technical Manual TM 38-760-1, *A Guide to System Engineering* (1973) and the Army Field Manual 770-78, *System Engineering* (1979). The EIA published SYSB-1, *System Engineering* (1989). Other standards followed, culminating in MIL-STD-499B, *Systems Engineering*, which was never formally adopted.

Eventually various professional societies joined together to publish ANSI/EIA 632 (1999), *Processes for the Engineering of a System*. The purpose of ANSI/EIA 632 was to create a standard which would be useful for both military and civilian applications. In addition, the Institute of Electrical and Electronic Engineering (IEEE) has published IEEE Standard 1220-2005, *Application and Management of the Systems Engineering Process* (2005). The standards, in general, pertain to product systems engineering (PSE) as defined in the SEBoK, described above.

The term SE is sometimes used in different contexts to mean different things, even within the commercial aircraft domain. For example, it can be used to mean the development of subsystems, as in the term *avionics systems engineering*. In addition, it is sometimes used to mean software engineering. This book uses the term in the broader sense which is consistent with the INCOSE definition discussed above and with the international standard ISO/IEC (2008).

SE is reaching a stage of maturity where its application in specific industries, such as commercial aircraft, can be defined and documented, as described by Petersen and Sutcliffe (1992). The first major guideline is the Society of Automotive Engineers (SAE) publication, ARP 4754 (1996), which applies the principles of SE to the development and certification of commercial aircraft. The later document ARP 4754A (2010) superseded the earlier guideline and contains many aspects of SE compared to the FAA manual with an emphasis on safety and certification.

1.4   Overview of this Book

The overriding SE principle stressed in this book is that the aircraft should be viewed as a whole and not as a collection of parts. Each chapter looks at a different aspect of SE and shows how that aspect would be reflected in the systems engineering of commercial aircraft. Chapter 2 looks at the commercial aircraft industry, describes the levels of aircraft development (new, derivative, and change-based) to which SE would be applied, and shows how the aircraft component architecture fits into the SE hierarchical model. This chapter also describes new technologies which would be applied to aircraft of the future. Chapter 3 introduces the SE concept of functions and shows how to apply functional analysis to the entire life-cycle of the aircraft, to the aircraft as a whole, and to the aircraft's subsystems. The SE concepts of performance requirements and constraints are the subject of Chapter 4. This chapter also shows how these requirements can be allocated to the aircraft's subsystems. Chapter 5 addresses constraints and specialty requirements and focuses on some key aircraft specialty areas, such as weight and reliability. The importance of the human factors aspect of cockpit design and the concept of organizational safety and its importance to the aircraft industry are also discussed in this chapter. Chapter 6 describes the SE concepts of functional and physical interfaces and shows how these concepts would apply to both external and internal aircraft interfaces. Chapter 7 shows how the SE concept of synthesis results in an actual aircraft design during the various stages of functional analysis, architecture development, and trade-offs. This chapter emphasizes the usefulness of quality function deployment (QFD) in this process. It also describes the process for introducing new technology into aircraft development. Chapter 8 shows how an aircraft is synthesized at the top level and how cost constraints are a vital factor in this process. Chapter 9 shows how the subsystems are synthesized and how this synthesis reflects the functions which have been allocated at the subsystem level. Chapter 10 describes how certification guidelines incorporate the SE philosophy and how safety analysis and software development fit into this philosophy. Chapter 11 stresses the importance of the verification of all requirements in aircraft development and shows how the SE verification concepts of test, demonstration, analysis, and examination fit into the verification of aircraft requirements. Finally, Chapter 12 discusses the key SE management and control activities, such as design reviews and configuration management, and emphasizes the importance of these activities to the aircraft development process. Chapter 13, first, provides a set of rules that will aid the commercial aircraft enterprise adapt the SE processes to that domain. In addition, Chapter 13 explains the roles that different organizations within the enterprise may play in the SE process. These include both technical and managerial organizations. Chapter 14 explains the commercial aircraft enterprise in a LSSI context with an emphasis on the supply chain. Chapter 15 explains the concept of risk management and discusses many risks that may be encountered in this domain, ways to anticipate them, and ways to mitigate them. Finally, Chapter 16 discusses the newly developing discipline of resilience and how it applies to the commercial aircraft domain.

1.5 Roadmap for Applying Systems Engineering to Commercial Aircraft

With these principles in mind, we can now show how the SE steps flow together for commercial aircraft development, as shown in Figure 1.2. This figure also provides the chapter and section numbers for a description of each step. As in Figure 1.1, this figure begins and ends with customer requirements and verification. But it then expands the core steps of functional and requirements analysis and synthesis to illustrate the following principles: (1) the flow down from top- to subsystem-level analysis, and (2) the treatment of new, existing, and certification-based SE.

By necessity, this figure is somewhat oversimplified. As discussed in Section 1.2, capturing customer requirements and developing system functions are two interrelated steps, both of which occur at the beginning of the SE process. Secondly, system synthesis occurs throughout the SE process. It is any step which leads to

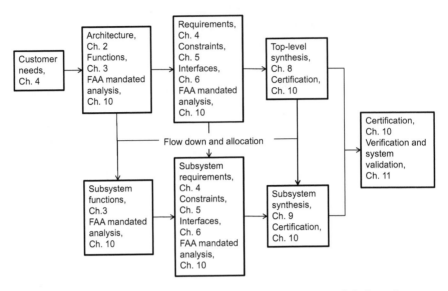

Figure 1.2   Roadmap to systems engineering for commercial aircraft

an aircraft design. Hence, both functional analysis and architectural development are part of this process.

1.6   Summary of Themes

This is not a systems engineering manual. As pointed out above, it is more of a *how* and a *why* guide rather than a *what* guide. If the reader wants a what guide, then authoritative sources, such as the *FAA Systems Engineering Manual* (2014), might be more appropriate. The guidance provided in this book is based more on an understanding of what should be done and the practical aspects of actually getting it done. With these thoughts in mind, let's look at some of the themes to be elaborated in later chapters. These themes are not in the order of importance; they are all important.

Theme 1—Adaptation

To execute all the processes in a manual such as the *FAA Systems Engineering Manual* might be perceived as an onerous and burdensome task. This book provides guidance on how these processes can be performed more efficiently without sacrificing quality. For example, Chapter 4 shows how requirements can be screened to determine which requirements are important and which ones can be ignored. The essence of requirements screening is risk, that is, the risk of ignoring requirements. Chapter 12 suggests ways to reduce the costly and time consuming process of design reviews. In short, all SE processes can be streamlined if appropriate attention is given to the risk of streamlining them.

Theme 2—SE as an integrated technical–managerial process

All systems engineering standards and textbooks involve the execution of tasks which can be considered both technical and managerial. This concept is counter to many conventional organizations in which SE is regarded as solely technical. Chapter 13 provides a tour of a typical organization containing both technical and managerial departments. This chapter explains the role of each department, both technical and managerial. This chapter also provides options regarding ways to bring the technical and managerial functions together in an integrated enterprise.

Theme 3—The importance of risk handling and management

No process permeates all aspects of an engineering organization than risk management. Any process that is performed imperfectly invites the specter of risk. No one has greater responsibility in this process than the program manager as explained in both Chapters 13 and 15. This is because the pressures to ignore or minimize the concentration on risks are enormous, and the consequences of doing so can be substantial. Also, as explained above in the discussion of Theme 1—Adaptation, risk plays an important role in that function.

Theme 4—The systems view

Essential to the discussion of systems engineering in any domain is an understanding of what a system is and the many manifestations of it. From the basic definition of a system in Section 1.1, one can see a system from many degrees of breadth, from the avionics system, to an entire aircraft system, to the entire air transport system. Chapter 14 also discusses the entire supply chain as a system of systems. Only with the systems view can one begin to answer the question of how each one of the systems functions as a whole and performs a useful function.

Theme 5—Added rigor

SE is not, as one might expect, an added layer of processes on top of the existing design processes. Some have described it as the glue that holds all the other processes together. In the Final Comments to the first edition of this book, it is described just as *common sense*. But in the end it must be concluded that SE is the process that adds rigor to the design, integration, and operation of a commercial aircraft.

2

Commercial Aircraft

Although the term *commercial aircraft* generally refers to jet-powered aircraft carrying large numbers of passengers for long distances, the SE principles outlined in this book also apply to freight-carrying aircraft and smaller propeller-driven, or *commuter*, aircraft as well. In addition, these principles also apply to *general aviation*, that is, small privately owned aircraft. Also included in commercial aircraft are regional jet aircraft which usually have fewer than 100 passengers and make trips internal to a country rather than internationally.

There are, indeed, many similarities to some classes of military aircraft which have missions of carrying passengers and cargo over specified distances. We show in Section 2.3, for example, that the ATA Specification 100 (1989) aircraft component hierarchy used in the commercial aircraft industry is also used in military practice (MIL-STD-1808B, 2007). The main difference is that the military hierarchy adds specific military categories, such as provisions to carry weapons. Other requirements unique to military aircraft include, for example, the need to provide protection from enemy weapons.

The main differences between commercial and military aircraft, however, lie in the types of requirements and the types of customers which generate the requirements. We will see in Section 3.1 that there are two types of commercial customers, the aircraft market and specific airline customers. While the military customer is normally a single customer with specific mission requirements, commercial aircraft requirements are strongly driven by economic requirements. In Chapter 8 we will discuss economic requirements as part of the top-level synthesis.

2.1   The Commercial Aircraft Industry

The commercial aircraft industry throughout the world is evolving to a small group of large companies. Some of the risks in this sector include the financial cost of new product development, and the financial health of airlines and other customers. Large established firms have been led to merge to accept the high costs and risks of doing business. This environment and the complex nature of aircraft development make commercial aircraft a prime subject for the application of SE.

2.2   Levels of SE Application

Aircraft development is conducted at three broad levels. Each level demands different aspects of SE. The levels are as follows:

Level 1—New aircraft

The development of new aircraft allows SE to be applied in a *blank slate* fashion: that is, to start from the inception of the requirements for an aircraft and the development of initial concepts. Discussions with *launch* customers are held, and analyses of range, number of passengers, noise, emissions, and other top-level requirements are developed. Economic, technical, and regulatory criteria are analyzed against all potential concepts. Requirements are allocated to the aircraft subsystems. Major components will remain the same for all customers except for those changed below as discussed in Section 2.2.

Level 2—Derivative aircraft

A derivative aircraft utilizes major components of existing aircraft as the basis for the development of an aircraft which meets some new requirements. The derivative aircraft may have increased performance or carry more or fewer passengers than the baseline aircraft. The challenge of SE is to develop requirements and to synthesize and verify solutions to those requirements within the specified constraints of the baseline aircraft. These are aircraft for which major components, primarily airframe, are used from previous models. These components will remain the same for all customers. The development of a derivative aircraft, as opposed to a new aircraft, can result in considerable savings in development and tooling costs, and lower prices for the customer.

Level 3—Change-based aircraft

A change-based aircraft is an aircraft for a specific customer which may have a large number of requested changes. Although each change may be small in itself, the SE methodology should be applied to each change to assure that the performance requirements of the affected subsystems are met and that the accumulated changes for the whole aircraft allow the aircraft to meet its own performance requirements. Individual customers provide requirements for specific changes, or the aircraft manufacturer initiates internally generated requirements for design improvements for economic or other reasons.

     Another idea frequently encountered in the industry is that derivative and change-based designs must meet less rigorous requirements than new aircraft. On the contrary, all aircraft are subject to *the top-level requirements.*

2.3 Aircraft Architecture

A key SE principle is that commercial aircraft should be considered *as a whole* and not as a collection of parts which can be independently developed and integrated. Requirements flow-down is dependent on viewing the aircraft architecture as a hierarchy in which lower-level elements, such as the subsystems, are subordinate to the aircraft. The aircraft itself is subordinate to a higher-level system called the aircraft system, which includes the aircraft and all its supporting systems.

The main reason for using the graphical depiction of an aircraft architecture is to illustrate its hierarchical nature, which corresponds to the hierarchy of the functions discussed in Chapter 3 and requirements discussed in Chapter 4. However, the hierarchy is not a hardware description; it is an abstract depiction of the aircraft. It is merely a set of *buckets* into which requirements can be placed. When the aircraft hierarchy is defined, one of the first steps, along with aircraft system-level functions discussed in Section 3.2, towards aircraft system synthesis will have begun. The hardware selection is the final step in system synthesis discussed in Chapter 7. Secondly, studying the aircraft architecture helps us understand what is really included in the aircraft *system*, which we will see below. This hierarchical architecture of aircraft fits within the SE model.

As we will see in Section 7.1, another principle is that the aircraft hierarchy is not permanent. As the design evolves, so will the hierarchy. Trade-offs may result in a new and improved hierarchy for a specific customer.

Concrete systems vs. abstract systems

Before we can discuss the aircraft hierarchy, it is necessary to understand the difference between a concrete system and an abstract system. In simple terms a concrete system is the system you can touch: You can touch the wheel; you can touch the wing, and so forth. An abstract system is a mental model, that is, it is a depiction of the system that is the aircraft, for the purpose of analysis. That is, an abstract system in anthropocentric view of a system. SE has adopted this hierarchical mental model as the primary method of depicting systems for analysis. This method is not totally random. It results from the fact that it is possible to view subsystems as clusters of components that exist at various levels of the abstract hierarchy. It is convenient that the traditional SE practice of viewing systems as hierarchies has also been adopted by the Air Transport Association of America (ATA) whose hierarchical depiction of an aircraft is shown below.

From an aircraft point of view, it is important to point out that there are other ways of depicting an aircraft system. Avionics specialists, for example, sometimes depict the avionics system as web rather than a hierarchy. This is because that system can be described at a single level rather than multiple levels. Neither method is correct nor incorrect. It is simply a matter of convenience.

The aircraft hierarchy and the Air Transport Association (ATA) index

The fact that the ATA has adopted this hierarchical view of the aircraft shows that the SE concept of an abstract hierarchy is already an accepted concept in the aircraft industry. Figure 2.1 shows a typical hierarchy of the aircraft system and its subordinate elements. This type of hierarchy is called a specification tree (or *spec tree* for short) since a specification should be written for each element as discussed in Section 12.6. The hierarchical breakdown of the aircraft system and the aircraft allows the flow-down of requirements to all subsystems and components in the classical SE manner.

In the ATA index the major aircraft divisions subsystems because they are not really subsystems; that is, they do not all perform a key function. Rather, they are collections of subsystems which group together technologically.

The arrangement of the ATA Specification 100 (1989) chapter numbers in Figure 2.1 is not a feature of the ATA index. Its purpose is to create a hierarchy useful for SE analysis and yet retain the ATA identification numbers. No components or ATA numbers are lost in this process. There is no requirement to break down the aircraft elements as Figure 2.1 does. There is also no requirement to make it correlate to the ATA index. However, doing so is a convenient way to follow the general SE practice of creating a hierarchical breakdown. Tying the breakdown to the ATA index creates a structure familiar to most aircraft engineers.

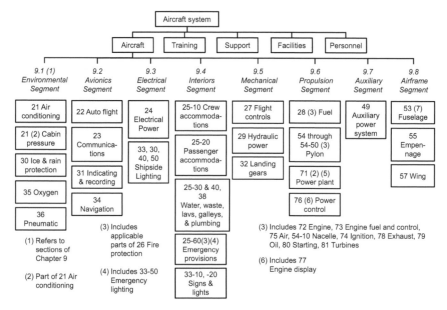

Figure 2.1   Generic aircraft system architecture and ATA chapter correlation

A goal of integrated product development (IPD) is to make the organizational structure and the aircraft hierarchy the same. The IPD process discussed in Section 12.3 divides the aircraft into *product centers*. Each product center is managed by an integrated product team (IPT). For example, a product center might be the wing. The wing IPT would be responsible for the entire development of the wing and all of the internal components. Hence, the program organization would be organized in accordance with the product breakdown. Thus, Figure 2.1 would represent both the product hierarchy and the program organization chart in which each box would represent an IPT at each organizational level. Alternatively, in an IPD environment the hierarchy shown in Figure 2.1 might be very different depending on the organizational structure chosen.

The aircraft system

The aircraft system consists of more than the aircraft (the flight vehicle) itself. The aircraft system consists of the following four elements, of which the aircraft is only one. Each of these subsystems can also be broken down into its subordinate components.

The aircraft The aircraft is the flight vehicle. The aircraft segments are described in Section 2.3.

Training equipment There are two types of training equipment normally included in training equipment. Flight crew training equipment includes simulators and any other equipment needed for flight training. Maintenance training equipment includes such equipment as mock-ups.

Support equipment There are two principal types of support equipment, on-aircraft and off-aircraft. Examples of off-aircraft support equipment are the ground electrical and hydraulic supplies. An example of an on-aircraft support equipment item is built-in test equipment (BITE).

Facilities The facilities of importance can either be buildings where aircraft are stored and maintained, or they can be specialized structures, such as the common airport skyways.

Personnel As discussed in Section 5.5, people can be considered to be a part of the aircraft system. It would be impractical, and perhaps unrealistic, to view people themselves as hierarchical systems. In a rudimentary fashion people might be considered to consist of a physical subelement and a cognitive subelement.

Aircraft segments

There are many ways to break down the aircraft into its subordinate elements. The following breakdown is typical:

1. *Environmental segment*—This segment includes air conditioning, ice and rain protection, cabin pressure, pneumatic supply, and oxygen supply equipment.

2. *Avionics segment*—This segment includes the communications, navigation, indicating and recording, and auto flight equipment. This segment might be more appropriately called the aircraft management segment since it includes the communications subsystem and the aircraft monitoring functions found in the indicating and recording subsystem.
3. *Electrical segment*—This segment includes electrical power and shipside lighting.
4. *Interiors segment*—This segment includes crew accommodations; passenger accommodations; water, waste, lavatories, galleys, and plumbing; emergency provisions; and interior signs and lights.
5. *Mechanical segment*—This segment includes landing gears, flight controls, hydraulic power, and cargo loading equipment.
6. *Propulsion segment*—This segment includes the engine pod and its components, fuel components, engine pylons, and thrust management equipment.
7. *Auxiliary segment*—This segment includes any auxiliary power supply, for example, for generating electrical or pneumatic power.
8. *Airframe segment*—This segment includes the wing, fuselage, and empennage.

Chapter 9 discusses the synthesis process as applied to each one of these segments.

Alternative hierarchies

As noted above, the hierarchy shown in Figure 2.1 is only one of many possible hierarchies. The development of an aircraft lends itself to many possible hierarchies. Following are two of the more obvious possible subsystems which might appear in an alternative hierarchy; each of these possible subsystems would contain portions or all of other subsystems shown in this figure:

1. *Cabin subsystem*—Although a cabin subsystem would be dominated by features of the passenger accommodations subsystem described in Section 9.4, it would contain portions or all of other subsystems. For example, it could also contain elements of the following subsystems: air conditioning described in Section 9.1, cabin pressure described in Section 9.1, oxygen described in Section 9.1, internal communications described in Section 9.2, electrical power described in Section 9.3, water, waste, lavatories, galleys, and plumbing described in Section 9.4, emergency provisions described in Section 9.4, and signs and lights described in Section 9.4.
2. *Cockpit subsystem*—In addition to all of the avionics segment subsystems described in Section 9.2, a cockpit subsystem would contain elements of air conditioning described in Section 9.1, cabin pressure described in Section 9.1, oxygen described in Section 9.1, electrical power described in Section 9.3, crew accommodations described in Section 9.4, water,

waste, lavatories, galleys, and plumbing described in Section 9.4, emergency provisions described in Section 9.4, signs and lights described in Section 9.4, flight controls described in Section 9.5, and power control described in Section 9.6.

2.4   Advanced Technologies on Aircraft

To meet design requirements for reduced weight, noise, and emissions, robust systems, and safe and economic operation, many advanced technologies are routinely incorporated into commercial aircraft, for example, heads-up displays (HUD), voice recognition, global positioning system (GPS) receivers, point-to-point inertial navigators, reconfigurable instrument displays based entirely on digital video displays, Doppler radar, fly-by-wire (FBW) or fly-by-light (FBL), and real-time computer fault detection and isolation. Composite material technology is key to weight reduction. Some of these technologies are discussed below, as described in part by Kehlet (1995). We will see in Section 7.6 how the SE process evaluates new technologies for incorporation into the aircraft design.

Advanced subsonic transports

For subsonic transports, key advanced technology applications include: center of gravity (c.g.) management systems, for example, with vertical stabilizer tanks; composite primary and secondary structures; supercritical wings with high-load alleviation, hybrid laminar flow control, and high-lift systems; advanced turbofan engines, FBW and power-by-wire (PBW); titanium landing gears; aluminum–lithium or metallic composite fuselage structures; and stability augmentation. Many, if not most, of these technologies are being incorporated into today's designs and will be routine in future aircraft.

Advanced supersonic transports

For supersonic transports, key advanced technology applications include synthetic vision, sidestick control, advanced lightweight materials, mixed flow turbofan engines, negative static margin, mixed compression inlets, arrow wing for supersonic cruise efficiency, FBL and PBW flight controls, and auto control in pitch. Technology requirements for advanced supersonic aircraft are discussed by *Aerospace Engineering* (1994) and Kehlet (1995).

Airframe technology

Improvements in the technology of aircraft structures will come from advanced materials and integration techniques. Integrated computer codes will allow the aerodynamics and strength aspects to be addressed simultaneously. Thus computer-aided design becomes an advanced tool for the evaluation of physical interfaces within the SE framework. The use of composite materials allows for both a decrease in weight and an increase in performance through higher-aspect ratios. Advanced machining techniques allow for the design to minimize parts.

According to Lin et al. (2013) "one of the main weaknesses of laminate composite structures" is delamination. Lin et al. provide an overview of techniques to alleviate this weakness. Although composite structures do not suffer from some of the weaknesses of metal structures, such as metal fatigue, delamination is a major weakness. Hence, before adopting this type of structure, the developer needs to address these techniques.

Aerodynamic improvements

New aerodynamic techniques include the use of pressure sensitive paint and computational fluid dynamics (CFD). These technologies will allow multipoint wing design that attains the lowest cruise drag characteristics and the highest realistic buffet onset boundary. Another goal is efficient aerodynamic profiles for wings with large high by-pass ratio (HBPR) engines. Another effort is the aerodynamically efficient but low-cost high lift systems. Studies have shown (Martínez-Val, 1994) that greater range and payload capability can be achieved with a third horizontal surface, or *canard*, located on the forward fuselage. Another concept of aircraft configuration is the blended wing-body (BWB), which resembles a large manta ray. This concept is described further below.

Noise control

Active noise control can reduce cabin noise without severe weight penalties. This technique introduces a secondary noise source of comparable amplitude but opposite in phase to the primary noise in order to cancel out the primary noise. It controls noise over a wide range of frequencies to counteract both engine noise and boundary layer noise. It will be especially important to control boundary layer noise on the high-speed civil transport (HSCT).

Fly-by-Light (FBL), Fly-by-Wire (FBW), and Power-by-Wire (PBW) technologies

FBL introduces multiplex photonically-based subsystems into the aircraft. FBL reduces wiring weight, reduces exposure to electromagnetic interference (EMI) hazards, and simplifies certification by eliminating the need for full aircraft subsystem tests. PBW and FBW result in weight savings and eliminate the need for engine bleed air and variable speed drives for secondary power subsystems. FBL, FBW, and PBW result in higher-reliability, lower-maintenance costs, and lighter weight.

FBW is an enabling technology for flight envelope protection as described below.

Synthetic vision capabilities

Synthetic vision enables pilots to use visual imagery and guidance cues to penetrate weather and compensate for low levels of illumination. These subsystems would use satellite-based navigation, imaging sensors, and high-resolution displays to operate with a greater degree of autonomy.

Propulsion controlled aircraft

An aircraft controlled by thrust modulation rather than control surfaces would be more able to survive catastrophic events, including terrorist actions, and perform better in partial failure conditions. Following the 1989 Sioux City DC-10 accident as previously described by Jackson (2010, pp. 78–79), there was much discussion of the possibility of propulsion control. In this incident the pilot was able to maintain partial control by using the propulsion control mechanisms. However, the NTSB (1990) did not recommend implementation of propulsion control but rather focused on other preventive measures.

Autonomous cargo handling

Improved methods for airlift cargo handling (IMACH) are an integrated group of technologies for improving cargo handling. These methods focus on handling functions, on the handling of large and more complex loads, and on automation features. IMACH can completely automate the movement of palletized loads. This improvement can achieve reductions in turnaround time, a prime cost driver for airline customers.

High Speed Civil Transport (HSCT)

The drive towards an HSCT has focused on many new advanced technologies. These technologies include advanced propulsion systems and advanced materials which can manage the temperatures associated with flight on the order of Mach 2.4. Research is being conducted in aerodynamics and technology integration, propulsion, structures and materials, flight deck systems, and key environmental issues, including sonic boom, airport and community noise, and emissions. Research has also begun on a hypersonic transport (HST) (*Aerospace Engineering*, 1996), for which the demands are even greater at Mach 5.0.

Human factors

Human factors have long been critical to aircraft design, especially in flight deck layout. Key demands include new techniques for solving the contradictory hazards of high work load and pilot boredom. Use of advanced visualization tools as part of the SE process in combination with design tools provides cost-effective, rapid prototyping to evaluate and adjust the design. We will see in Section 5.5 how SE integrates human factors into the requirements process.

Advanced design tools

The complexity of aircraft design led the aircraft industry to develop sophisticated design and simulation tools. In many cases such tools provide design information faster and more cheaply than total reliance on wind tunnel testing. These tools, combined with advanced visualization techniques, are sufficiently mature to provide a basis for research as well as aircraft design and engineering. There is a strong synergism between the computer-aided design tools and the training for each aircraft type. The modeling of flight characteristics for new aircraft types is so accurate that type checkout for pilots based only on trainer experience is anticipated. Simulation has been established as one of the primary SE verification techniques in the commercial aircraft industry.

Flight envelope protection

In recent years a new system has emerged called *flight envelope protection*. According to Airbus,

> Fly-by-wire [the use of digital rather than mechanical interfaces] enhances safety by allowing the programming of the flight envelope protection, which enables pilots to fly the aircraft freely but prevents any abnormal operations, such as stalling, flying too fast, or overstressing. (Airbus, 2013)

Other manufacturers have favored other approaches to safety. To date there is no universal agreement as to which is the most effective approach. However, as seen in Section 10.4, the international safety consortium CAST has recommended that some aspects of flight envelope protection be implemented on all new aircraft.

Blended-Wing-Body (BWB)

Originally conceived by McDonnell Douglas prior to its merger with Boeing in 1997, the blended-wing-body is a concept with much promise for carrying more passengers or cargo more economically. It is similar to a flying wing except that it consists of a wing and fuselage that blend together smoothly. The BWB showed advantages over conventional aircraft in operating costs, fuel efficiency, gross weight, and nitrous oxide emissions.

In spite of its promise, the BWB still has many challenges. In a NASA report Bowers (2000) provides a comprehensive summary of both the advantages and disadvantages of this concept. One disadvantage is that the plane would be too large for current gates. Folding wings were investigated and determined to be unacceptable. Other challenges included structures and materials, aero-structural integration, aerodynamics, controls, aero-structural integration, propulsion-airframe integration, systems integration, and infrastructure.

At the time of his report, Bowers states that NASA was still committed to supporting research on the BWB. For its part, Boeing has not announced any intention to introduce this concept into commercial practice.

2.5 Aircraft Manufacturing Processes

The aircraft industry has in recent years seen a shift from virtually hand-made to mass produced aircraft. Although many steps in the manufacturing process of commercial aircraft parallel those for motor vehicles, design and manufacture of commercial aircraft are even more complex. Although aircraft manufacturing is a low-volume process, its complexity arises from dependency on highly integrated high-technology subsystems, use of advanced materials, detailed specifications, and extremely rigorous testing. SE is especially suited to addressing these issues.

2.6 Trends in Commercial Aviation

Economic and regulatory pressures

Pressures in the commercial aircraft domain come from all directions, both economic and regulatory. Economic pressures come primarily from the airline customers. Of course, every airline wants to carry as many passengers as far as possible and as cheaply as possible. These pressures have resulted in more efficient engines and radically different designs, such as the blended-wing-body discussed above. Another trend is toward more efficient flight paths. One concept being looked into is the 4D trajectory. According to Skybrary (2014),

> The 4D trajectory of an aircraft consists of the three spatial dimensions plus time as a fourth dimension. This means that any delay is in fact a distortion of the trajectory as much as a level change or a change of the horizontal position. Tactical interventions by air traffic controllers rarely take into account the effect on the trajectory as a whole due to the relatively short look-ahead time (in the order of 20 minutes or so).

Among the many benefits of the 4D trajectory are optimal operations for airlines, reduced cost and time, reduced emissions, and reduced load on controllers.

Other pressures especially on engine makers are the pressures to reduce carbon footprint and engine noise.

Trends in component procurement

A recent trend in component procurement is for airline companies to procure or lease individual components themselves rather than using the supplier-provided components specified by the aircraft developer. Although the economic benefits for the aircraft operator may be substantial, there may be risk associated with this trend if the procured parts are not flight qualified. Parts may be valves, pumps, fans, and other parts. First of all, the environments for these parts may vary widely from airline to airline, from aircraft to aircraft and from aircraft zone to zone. The environment in an engine nacelle is considerably hotter than, for example, in the cargo bay. Hence, the responsibility is on the aircraft operator to assure that all the parts procured in these ways are flight qualified.

3

Functional Analysis

… a motorcycle can be divided according to its components and according to its functions..The overall name of these interrelated structures, the genus of which the hierarchy of containment and structure of causation are just species, is system. The motorcycle is a system. (Robert Pirsig)

The International Council on Systems Engineering (INCOSE) Fellows (2006) state that a basic property of a system is its function. A function is a task, action, or activity performed to achieve a desired outcome. This principle applies to any system, including a motorcycle, as explained above by Pirsig (1974). This chapter will show how the functional architecture of the aircraft system becomes the basis for all aircraft SE analyses. It will show that the functional architecture is, to a given level, valid for all aircraft.

Functional analysis is a critical part of the SE process for commercial aircraft for three reasons:

- First, functional analysis is one of the primary techniques for establishing the completeness of performance requirements as discussed in Section 4.2. Functions are the primary prerequisite for the establishment of all performance requirements. This relationship between functions and performance requirements is valid for all systems, not just aircraft systems.
- Secondly, guidelines such as ARP 4754A (2010) for the certification of commercial aircraft call for a functional hazard assessment (FHA) based on a comprehensive functional analysis of the entire aircraft as shown in Section 10.2. The FHA examines all aircraft-level and subsystem-level functions and determines the safety criticality of these functions during aircraft operation.
- Finally, the functional architecture is the basis for the architecture of the entire aircraft. According to Rechtin (1991, p. 212), "Except for good and sufficient reasons, functional and physical structuring should match." This rule applies both to the top-level architecture of the system and also to the architecture of individual subsystems. The Advanced Design organization can use this rule to create the architecture of the aircraft and of the individual subsystems.

The development of the functions of the aircraft system as discussed in Section 3.2 is one of the first steps in the aircraft system synthesis process, for it is around the system functions that the nature of the design begins to take shape. The final step in system synthesis occurs when the functions and requirements are converted into a design, as discussed in Chapter 7.

A sound practice of functional analysis is to define the limits of each function. These limits are normally defined by time and the scope of the activities included in the function. That is, the function has a definite beginning and end, and definite activities, all of which should be known. The importance of this concept will become more apparent for flight operations as discussed in Section 3.3.

Throughout this book, the names of functions will be capitalized. The names and organizations of functions are, to a certain extent, dependent on the interpretation of the analyst. So the reader should not interpret the functions and diagrams in this chapter to be definitive. It is the purpose of this chapter primarily to illustrate the significance and usefulness of functions and functional analysis.

When to perform functional analysis

Given the considerable effort that functional analysis may require, it is appropriate to ask: when should functional analysis be performed and when can it be skipped? Chapter 2 defines three types of commercial aircraft: new, derivative, and change-based. The most appropriate time to perform functional analysis is for new aircraft. This is because the aircraft is being created from scratch: its architecture, components, and requirements. For derivative and change-based aircraft the architecture and the components will, for the most part, already have been defined. For these cases, functional analysis can be avoided, or at a minimum modified only slightly. But, of course, the FHA will need to be done for all three types of aircraft. In addition, it must be remembered that the performance requirements may not have been defined for these cases. Or, perhaps the newly derived aircraft or changed aircraft may result in new performance requirements. For this reason the performance requirements should be revisited for these components. See Chapter 4 for a complete discussion of performance requirements and their development.

The creation of a function

Before we discuss actual functions, we need to ask the questions: what is a function and how do you create one? A function is a description of what a system element does, but the system element is not named in the function. When you name the system element, you have a performance requirement. Without the system element, you just have a functional requirement.

The simplest form of a function is very simple; it is just a verb and a noun. Typical functions are *generate power* and *provide lift*. It is pretty obvious that a generator will generate power and a wing will provide lift, but that is not part of the function. Some practitioners try to avoid the verb *provide* in the name of the function, but logically there is nothing wrong with it.

The logic behind not naming the element is that this approach leaves the designer open to alternative ways to satisfy the function.

3.1 The SE Life-Cycle Functions

Table 3.1 compares the aircraft life-cycle functions with the traditional SE life-cycle functions from ANSI/EIA 632 (1999) and the life-cycle functions from the *FAA Systems Engineering Manual* (2014). The aircraft life-cycle functions have been organized to emphasize various aspects of the commercial aircraft life-cycle. These are the activities performed by the *developer* and the *users* of the aircraft from the moment of its conception to its disposal. That is to say, they are not the functions performed by the aircraft itself. Those are the functions of primary interest, but we will discuss them later in Section 3.3.

It is important to discuss the life-cycle functions because they have a significant impact on the aircraft itself. Furthermore, the life-cycle functions include the entire SE process, which is the subject of this book.

Figure 3.1 shows a typical SE life-cycle functional flow[1] for the development, manufacture, operation, and disposal of an aircraft system. This flow incorporates the aircraft life-cycle functions of Table 3.1. The table compares these phases with the traditional SE life-cycle phases and the FAA (2014, p. 5) life-cycle phases. This comparison shows that the different views of life-cycle phases only differ by scope.

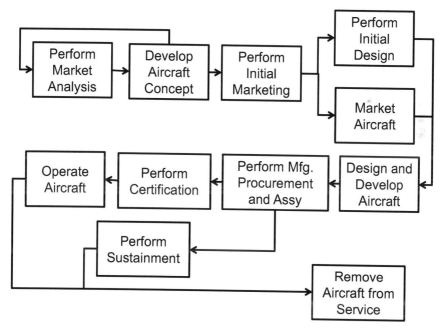

Figure 3.1   SE life-cycle functional flow for a commercial aircraft system

---

1   A rigorous functional flow diagram will have *and* and *or* gates. These have been eliminated for simplicity.

The important thing to remember about life-cycle functions is that they define the functioning of the entire aircraft system beyond the aircraft itself, and since this system is a *system*, all the pieces of it will relate to each other and not just constitute independent and unrelated components.

Table 3.1　　Comparison of life-cycle functions

Traditional SE Life-Cycle Functions	Aircraft Life-Cycle Functions	FAA Life-Cycle Functions[*]
Development	Market analysis Perform initial marketing Perform initial design Market aircraft Perform design and development	Mission analysis Investment analysis
Manufacturing	Perform manufacturing, procurement, and assembly	Solution implementation
Verification	Perform design and development Perform certification	
Deployment	Operate aircraft	In-service management
Operations		
Support	Perform sustainment	
Training		
Disposal	Remove aircraft from service	Disposal

[*] *Source: FAA Systems Engineering Manual* (2014, p. 5).

The Perform Market Analysis function

The SE process begins, and the aircraft concept is born during this phase. This is a subfunction of the Development function in Table 3.1 and is where the basic top-level requirements are established. The process of identifying top-level requirements begins by identifying the types of customers. There are two kinds of customers: the aircraft market and specific customers. This function addresses only the general market as a customer. The basic questions to be answered are: How large is the market? How many seats are optimum? What are the airline route structures which will establish the range requirements? Are there any special requirements, such as short runways to consider, and so forth? However, the most important questions to be answered during this phase are economic ones: How many passengers are anticipated to fly in the near future? What will be the direct operating cost (DOC) of the aircraft the potential customers will expect as discussed in Section 8.6? What is the maximum price potential customer will be willing to pay for an aircraft? The answers to all of these questions will influence the design.

The Develop Aircraft Concept function

Within this function, another subfunction of the Development SE function, the systems engineer creates a concept, using the top-level requirements from the market analysis. The concept of this phase is what it is, namely, *a concept*. It is not a detailed design. Only engineering sketches will be produced. This phase will reveal such parameters as the total estimated weight, the number of engines, and so on. But detailed information will not be known. The purpose of the concept is to verify that an aircraft can be designed and built to meet the market requirements. Concept development involves both the establishment of the initial aircraft architecture as discussed in Section 2.3 and the top-level aircraft system functions shown in Section 3.2. The concept of this phase is *subject to change* in the subsequent phases. After all, we have not yet determined specific customer expectations. The resulting aircraft may be very different indeed from the concept of this phase.

The concept developed in this phase may, of course, not even be a new aircraft as described in Section 2.2. It may be a derivative aircraft also described in Section 2.2, or it may be a change request-based design also discussed in Section 2.2. In any case, the adherence to SE principles will apply.

The Perform Initial Marketing function

Here is where we find out what specific customers' requirements are. Market analyses discussed in Section 3.1 will determine the need for aircraft in specific range and payload classes. For specific customers, the most effective way to perform this process, from an SE point of view, is to determine customer expectations from a functional point of view. That is, rather than emphasizing hardware options, we can stress the functions the customer wants to see. One way to do this is quality function deployment (QFD) described in Section 7.4. QFD is a very effective process for determining and prioritizing customer needs and presenting concepts to the customer which meet those needs. Based on the results of the real top-level requirements and an analysis to verify that the requirements are achievable, we are now ready to conduct a system requirements review (SRR) of the aircraft described in Section 12.4. The SRR provides concurrence from the customer(s) that the requirements are correct. Section 8.1 presents a list of typical top-level requirements which can be determined from market analyses and from specific customers.

The Perform Initial Design function

Based on the real requirements established with one or more prospective customers, we can now create a concept for a real aircraft. This step is also part of the Development function of Table 3.1. We will have established the basic characteristics of the aircraft but only at the top level. We are now ready to conduct a system design review (SDR) described in Section 12.4. This review, also with customer(s) present, confirms that the concept meets the requirements. The completion of the SDR signals the beginning of initial design and, hence, the transition from concept to design. The result of a successful SDR is an approved aircraft system specification.

One of the functions of an SDR is to identify the trade-offs that need to be performed before the next review, the preliminary design review (PDR) described in Section 12.4. System trade-offs are one of the main functions of the synthesis process described in Section 7.2. Here is where we evaluate the introduction of new technologies, such as composite materials. We have seen in Section 2.4 some of the possible technologies which may be introduced.

Chapter 12 discusses the various ways that these design reviews can be modified to adapt to the commercial aircraft domain without sacrificing quality.

The Market Aircraft function

In the commercial aircraft industry, marketing is much more than selling. Marketing is also part of the Development function of Table 3.1. From an SE point of view, the key marketing function is the determination of customer requirements. Now that we have settled on a basic aircraft design in the initial design phase above, the requirements for future marketing focus will be all change-based discussed in Section 2.2. That is, each customer will sometimes require extensive custom features on the fleet of aircraft. As we said before, these requirements have a significant impact on the aircraft design and therefore deserve as much attention and rigor as the aircraft-level requirements. With each custom feature comes the need for maintaining the aircraft performance, reliability, and safety, and meeting the host of other constraints. We can and should also perform QFD described in Section 7.4 with each of the new customers.

The Perform Design and Development function

In this phase we develop the detailed requirements described in Chapter 4 and conduct PDRs described in Section 12.4. We produce the detail drawings and review them at Critical Design Reviews (CDRs) also described in Section 12.5. During this function, verifications at the subsystem level are performed. Hence, this function includes both the Development and Verification functions of Table 3.1.

A key aspect of this phase is not only to design and develop the aircraft but also to develop the requirements for the manufacturing processes to follow. This is another example of the thoroughness of the SE process and its ability to determine the requirements for all aspects of the system far beyond the design of the aircraft itself.

As we will see in Section 10.2, the concept of organizational safety comes to the fore within this function. This concept requires that organizational aspects, such as aggressive schedules, do not increase the likelihood that the aircraft will have hazardous features.

The Perform Manufacturing, Procurement, and Assembly function

The Manufacturing function of Table 3.1 is an example of a phase of the SE process, other than Operate Aircraft, which can assign requirements to the aircraft itself. The most obvious example is the cost constraint imposed by the recurring (unit) cost of the aircraft as discussed in Section 8.6. A major driver of cost constraints is the cost of assembly. For example, this cost drives the number of parts to a minimum. Another example is the requirement to transport the aircraft to a central assembly facility in parts small enough to carry on a truck as discussed in Section 5.12. This requirement will force the design of the aircraft in segments rather than in large assemblies.

The Perform Certification function

After we assemble the complete test model, we flight test it and certify it. Flight test and certification are also part of the verification process described in Section 11.2.

The Operate Aircraft function

This function includes both the Deployment and Operations functions of Table 3.1. This is the phase that receives the lion's share of attention during the initial requirements definition phase of the aircraft. It is for this phase that we establish the top-level requirements based on the customer needs and determine them for the subsystems described in Sections 8.2 and Chapter 9. It is here that most of the requirements for the aircraft system are developed as discussed in Section 3.2. However, it is important not to forget the other phases.

The Perform Sustainment function

Sustainment is one of the most obvious, yet ignored, phases of the SE life-cycle. Sustainment includes both the Support and Training functions of Table 3.1. Maintainability engineers make valiant attempts to have their concerns addressed in the aircraft design. But traditional processes do not make that easy as discussed in Section 5.7. Only the SE process recommends the consideration of this phase as part of the functional analysis and the subsequent requirements development. Of course, maintainability is not the only sustainment function. There are also servicing and training, to name two. All of these provide aircraft requirements.

The Remove Aircraft from Service function

The need for disposal creates requirements for the aircraft. For example, regulatory documents prohibit the use of toxic materials. This function corresponds to the Disposal function of Table 3.1. This is another example of the impact of the life-cycle requirements on the aircraft.

3.2   Aircraft System-Level Functions

We are free to define the aircraft system as broadly as we need to. For example, we could define it to include the entire air transport system of the world, including all support systems and traffic control systems. For our purposes we only need to define it to include the aircraft, the pilot, and all the support equipment, training equipment, and facilities for the aircraft. We can call this the operational equipment, as opposed to business support equipment.

There are two reasons to identify the aircraft system functions and aircraft functions *at the same time*: First, the aircraft concept may require changes in some of the training, support, and facilities used with the aircraft. Certainly, with the introduction of major aircraft changes, such as the High-Speed Civil Transport (HSCT), this statement will be true. Secondly, the interfaces with these other non-aircraft elements may affect the design of the aircraft itself. Historically, aircraft designers have either ignored these aspects or postponed them until later. Most of the functions for the aircraft system, other than the aircraft itself, will be subfunctions of the Perform Sustainment function described in Section 3.1.

Training functions

Typical training functions include Provide Training for Flight Crews and Provide Training for Support Crews. The Provide Training for Flight Crews function may result in new simulators. More importantly, from an aircraft point of view, the aircraft itself may need special training features to facilitate the training function. It is a basic principle of SE that we can allocate requirements to any element. Hence, we can allocate the requirements resulting from the Provide Training for Pilots to the aircraft element. The same is true of any function.

Support functions

Typical support functions include Provide On-Aircraft Support and Provide Off-Aircraft Support. The support functions are probably the most important aircraft system functions except for the aircraft itself. They are very broad functions and cover many activities, including maintenance, servicing, and the replenishment of provisions for the crew and passengers. The maintenance functions will determine whether special maintenance tools or other equipment is needed, such as test equipment. These functions will influence the design of cargo loading equipment and other service equipment. If a more common environment exists in which the customer equipment is fixed, then we should consider these items as fixed external interfaces and design the aircraft to satisfy them. Consideration of the fixed interfaces should be part of the design process.

One of the most costly support functions of the airline is the acquisition, storage, and disposition of spares. The number of spares is a strong function of the airline's route structure. That is, the airline normally chooses to store spares at key *hub* facilities, which strongly affect the number of spares. Hence, the systems engineer should consider the cost of spares in the total system cost of the aircraft.

Facilities functions

Typical facilities functions include Provide Facilities for Storage and Maintenance of Aircraft and Provide Passenger Access to Aircraft. The facilities functions will determine whether new or modified facilities are needed to store and maintain the aircraft. Certainly, radically changed aircraft dimensions would influence these decisions. In addition, the system requirements may alter the passenger access tunnel, called the jetway. We will have to take into account the effect of the jetway on the aircraft itself, since the jetway interfaces directly with the aircraft. For example, external objects, such as pitot tubes cannot be located in the vicinity of the jetway attachment to the aircraft.

3.3 Aircraft-Level Functions

The concept of operations functional view

Finally, we get to the aircraft itself. The general functional description of how the aircraft operates is called the Concept of Operations (CONOPS), as shown in Figure 3.2 adapted from NASA (2001). This figure presents the functions of the aircraft from an operational point of view.

Future diagrams will employ the functional flow block diagram (FFBD) which is the preferred format of the *FAA Systems Engineering Manual* (2014, Section

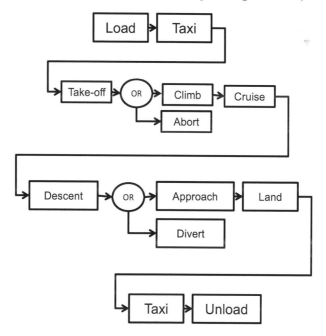

Figure 3.2   The commercial aircraft concept of operations

4.4). However, many analysts prefer the IDEF0 (Integrated Definition for Function Modeling) methodology. The advantage of IDEF0 is that it accounts for the inputs and outputs of each function. This method will be shown also. There are other functional diagrammatic methods, such as the swim lane diagram also shown. The advantage of this method is that it emphasizes the time relationship among the components of the system.

Table 3.2 compares the phases of operation with the aircraft-level functions. However, this table does not show all top-level functions. This view only shows this function from an operational phase point of view. In reality, the subfunctions of the CONOPS are multidimensional.

Table 3.2   Comparison of aircraft phases of operation and aircraft-level functions

Phases of operation	Aircraft-level functions
Load	
Taxi	Provide Ground Movement (pre-flight)
Take-off	Perform Flight Operations
Climb or Abort	
Cruise	
Descent	
Approach or Divert	
Land	
Taxi	Provide Ground Movement (post-flight)
Unload	

Each of the flight functions shown in Figure 3.2 should be matrixed against three other functional dimensions, as shown in Figure 3.3. The large number of functions implied by this chart, especially when expanded to the subsystem level, illustrates the monumental task required to assure the aircraft safety when the functional analyses suggested by the FAA in 4754A (2010) are developed.

All combinations of functions in this matrix do not necessarily exist. For example, the combination of the Perform Pre-Flight Operations function and the Provide Aerodynamic Performance function is not relevant because the aircraft does not fly during the pre-flight phase.

The concept of defining the functions precisely, as discussed at the beginning of this chapter, is especially important when considering the flight functions. The failure to define the limits may result in incorrect requirements and inadequate solutions.

Let's discuss each of the nodes of the Perform Air Transport Mission matrix:

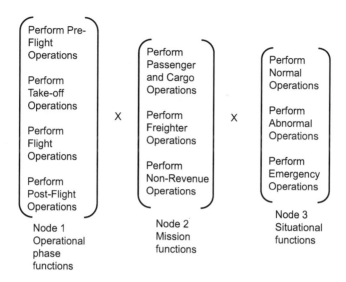

Figure 3.3   Matrix of the Perform Air Transport Mission function

Node 1: Operational Phase functions

We have broken the flight phases down into four basic phases shown in Figure 3.3. These are the same basic phases which define the Perform Air Transport Mission function of Figure 3.2 However, it may (and probably will) be necessary to break down the flight phases into even smaller phases during the analysis of specific subsystems. The flight phases which are important to the environmental control system (ECS), for example, will be different from the phases important to the flaps. The definition of the external environment associated with each phase is an essential aspect of the flight phase functions.

The Perform Pre-Flight Operations function
This function begins when the aircraft has begun to be prepared for flight. This function is related to the Perform Sustainment Function described in Section 3.1 of Figure 3.1 in that it consists of all the servicing and maintenance activities performed when the aircraft is either in a maintenance facility or on the airport ramp. The maintenance functions performed in these locations do affect the aircraft design. One such function would be Allow Removal and Replacement Access.

The Perform Take-Off Preparations function
This function begins when the aircraft pushes away from the gate and ends when it receives clearance for take-off. Requirements pertaining to runway transit, aircraft navigation, and communications will be established during this phase.

The Provide Ground Movement function, shown in Figure 3.4, is a subordinate function to the Perform Take-Off Preparations function. In addition, most of the subfunctions of the Perform Flight Operations function described in Section 3.3 will apply. The primary subsystem to implement the requirements of this function is the landing gear and brakes subsystem described in Section 9.5. The landing gear subsystem will provide the carriage and steering while the aircraft is on the ground, and the braking subsystem will provide the braking.

The Perform Flight Operations function

This function begins when the aircraft receives clearance for take-off and ends when it pulls off the runway. Obviously, many subfunctions will occur between these times, for example, take-off, climb, cruise, descent, approach, and landing. Each of these subfunctions is important in aircraft sizing described in Section 8.2. Additional subfunctions can be defined which will lead to requirements definition for various subsystems. For example, the raising and lowering of flaps will define functions important to the flaps and other subsystems, such as hydraulics. We will expand the flight functions later in Section 3.3.

The Perform Post-Landing Operations function

This function begins when the aircraft turns off the runway and ends when it arrives at the gate. The requirements defined during this phase will be similar to those identified in the take-off preparations phase described in Section 3.3. This function will include the Provide Carriage, Braking, and Steering subfunction as shown in Figure 3.4 as well as most of the subfunctions of the Perform Flight Operations function described in Section 3. However, the possibility that pre-flight and post-flight functions may be different compels us to examine them separately.

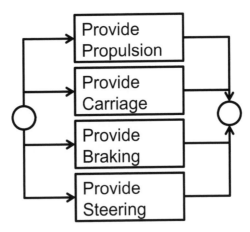

Figure 3.4   The Perform Ground Movement function

Node 2: Mission functions

The three mission functions shown in Figure 3.3 may apply separately or in combination.

The Perform Passenger and Cargo Operations function
This is the normal airline functional mode. It includes all the operations necessary to transport passengers and cargo between airline facilities.

The Perform Freighter Operations function
This function is well-known, for many airlines operate aircraft with the sole purpose of transporting freight. However, in the event the customer desires an aircraft with the capability of being converted from a passenger and cargo configuration to a freighter configuration, this function should be considered *in combination with* the Perform Passenger and Cargo Operations function. The reason for this requirement is that the convertibility requirement may impose special features on the aircraft that a pure passenger and cargo or a pure freighter version would not have.

The Perform Non-Revenue Operations function
All aircraft are required, from time to time, to operate in a non-revenue mode, that is to be flown from one point to the other with no passengers or freight. This function accounts for any requirements which may result from this mode.

Node 3: Situational functions

This group of functions reflects the functional operations of the aircraft in a variety of situations related to the safety of the aircraft and the occupants.

The Perform Normal Operations function
This is the basic functional mode in which the aircraft operates as expected through all phases of flight.

The Perform Abnormal Operations function
This function is the basic operational mode in which the aircraft or some subsystem of the aircraft is not operating normally, but no danger to the passengers or crew exists. For example, if a generator malfunctions and a backup power system is providing sufficient power, we would say that the aircraft is performing the Perform Abnormal Operations function. This function is important in establishing the requirements for backup systems.

The Perform Emergency Operations function
This is the functional mode in which a possible catastrophic event is likely. It is important in establishing the requirements to survive the expected environment.

This function is also responsible for the protection of passengers in emergency conditions, such as the loss of cabin pressure, or in emergency landing operations. This function also leads to the requirement for the emergency equipment and features which should be carried on each aircraft.

Expansion of Flight Operations functions

We saw before in Section 3.3 that the Perform Flight Operations is just one of four operational phase functions. Figure 3.5 shows the subfunctions of the Perform Flight Operations function. These subfunctions are the heart of the architecting of a commercial aircraft. The key point to remember here is that functions are only functions. They are not subsystems. That is, we cannot assume that the Provide Environmental Control function only belongs to the Environmental Control Segment or that the Provide Passenger and Crew Accommodations function only belongs to the Interiors Segment. On the contrary, each of the following functions may be allocated to several of the aircraft segments.

It is this ability to trade off higher-level functions among several segments which allows optimization of the system (the aircraft) at the aircraft level rather than at the segment or subsystem level. Failure to optimize at the aircraft level may lead to optimization at the subsystem level and hence to a system which is not optimum.

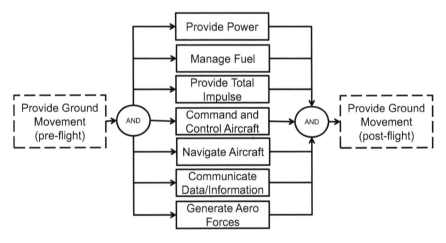

Figure 3.5   The Perform Flight Operations function

Although the flight functions listed in the following paragraphs may appear to pertain to individual subsystems, the SE process does not allow us to make that judgment at this time. They are still *top-level* functions. Not until Chapter 9 do we address the subsystems associated with the performance requirements of these functions.

The Perform Flight Operations function (IDEF0 view)

Originally developed as a software modeling tool, IDEF0 (Integrated Definition for Function Modeling) is widely used in SE as discussed by Buede (2000). The advantage of IDEF0 over the FFBD is that IDEF0 identifies the inputs and outputs of each function. FFBD focuses on the sequential occurrence of functions. These inputs and outputs form the basis for the cluster analysis to be shown at the end of this chapter. IDEF0 normally requires the assignment of these functions to system elements (components and subsystems), but for simplicity these elements are not shown in Figure 3.6. Figure 3.6 presents the same functions as in Figure 3.5 but in the IDEF0 format. The assignment of these functions to subsystems will be shown within the individual box describing each function.

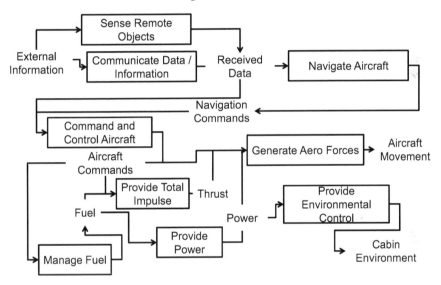

Figure 3.6   The Perform Flight Operations function (IDEF0 view)

The Generate Aero Forces function
We can break down the Generate Aero [aerodynamic] Forces into the following major subcategories: Provide Lift Performance, Provide Drag Performance, Provide Aerodynamic Stability, and Provide Aerodynamic Control, as shown in Figure 3.6. The primary aircraft segment allocated to perform the requirements associated with this function is the airframe segment described in Section 9.8. Of course, other segments, such as electrical described in Section 9.3 and mechanical described in Section 9.5, are strongly affected since they are required to move the control surfaces.

Of course, the lift and drag *performance* is the primary factor which allows the aircraft to achieve its speed (cruise, landing, take-off), range, cruise altitude, and other top-level performance requirements. These functions are achieved through the aerodynamic shaping of the wing and various other external components. Most manufacturers are employing winglets to enhance the lift and drag performance. Flaps provide augmented lift and drag for landing. The airframe segment described in Section 9.8 bears the major responsibility for satisfying these functions.

The aircraft achieves aerodynamic stability through the lift and weight distribution if the *static margin* is positive; that is, that the lift forces act through a point (the center of pressure) a fixed distance behind the center of gravity. The tail surfaces (the empennage) provide the aft lift forces to assure that this is true.

Many aerodynamic surfaces provide control. These surfaces include the rudder (horizontal control), elevator (vertical control), and ailerons (roll control). In addition to these surfaces, the spoilers aid both the flaps and ailerons for landing and roll control.

The primary inputs to this function are the aircraft flight commands and power for moving the control surfaces.

The Provide Total Impulse function

This is the primary function for providing the propulsive force called thrust that propels the aircraft throughout its mission. The central subsystem for this function is the propulsion system.

The inputs for this function are fuel and aircraft commands to control the proper amount of thrust.

The basic subfunctions of Provide Total Impulse are Provide Forward Thrust and Provide Reverse Thrust. Subordinate functions will include Start Engines and Shut Down Engines.

Most modern commercial aircraft employ turbojet engines for thrust. However, there are a significant number of turboprop and reciprocating engine aircraft in service. We will see below and in the propulsion segment discussion discussed in Section 9.6 that the engines normally provide a number of functions other than providing thrust. These include Provide Bleed Air and Provide Mechanical Torque (for electrical generation). This example illustrates the generally held principle that a single subsystem can have performance requirements from multiple functions.

The Manage Fuel function

This function governs how much fuel is provided to the engines. In addition most aircraft have multiple fuel tanks, usually in the wings, to allow the fuel to be transferred from tank to tank in order to control the c.g. of the aircraft. The element of the aircraft that performs this function is the fuel management system, normally a separate software module within the aircraft.

Inputs to this function are from the aircraft control commands provided by both the pilot and the automated control system.

The Command and Control Aircraft function

This function includes all commands that are necessary to move control systems or perform other actions to control the aircraft during flight including managing the fuel system and the propulsion system.

Two subsystems perform this function. One is the pilot and the other is the aircraft control system. That is, from an SE point of view, the pilot is a legitimate subsystem; otherwise the aircraft would not be a complete system. For two subsystems to perform this function there needs to be a complete set of rules to manage the interaction between the pilot and the automated control system. Billings (1997, pp. 237–260) has compiled a list of these rules. Among these he states, "To command effectively, the human operator must be involved."

Inputs to this function are first commands received directly from the navigation system and secondly information received directly by the pilot from information both inside the aircraft, the flight deck, and outside.

The Navigate Aircraft function
This function has several basic subfunctions: Determine Location of Aircraft, Determine Attitude of Aircraft, Determine Speed of Aircraft, Determine Direction of Aircraft, and Provide Flight Management. The majority of the requirements of these functions are the responsibility of the avionics segment described in Section 9.2.

Location can have many meanings. It includes the altitude, latitude, longitude, and the location relative to specific ground points. The attitude includes the pitch, yaw, and roll angles relative to an absolute coordinate system and to the air. The speed and direction can be relative to the air and to the ground. Speed also includes the rate of climb or descent. The Command and Control Aircraft is the function which provides commands to both the pilot and to the autopilot to control the aircraft and receives commands from this function.

The Communicate Data/Information function
There are two basic types of communications as expressed by the subfunctions: Provide External Communications and Provide Internal Communications. The requirements associated with these functions are the primary responsibility of the communications subsystem described in Section 9.2. External communications provide the means for the pilot to exchange information with the tower, with other aircraft, and with other ground nodes.

In addition, the pilot and other crew members must be able to talk to each other and to the passengers. Also, there must be a communications link between the flight deck and various parts of the aircraft for communication with the ground crew when the aircraft is on the ground.

This link supports the basic function of providing the pilots with sufficient information to fly the aircraft safely and monitoring the status of aircraft subsystems for malfunctions and any other hazardous condition. In addition, it provides the pilot with information outside the aircraft. Section 5.5 discusses the human factors aspect of establishing and implementing the requirements associated with this function. The indicating and recording subsystem described in Section 9.2 provides most of the information associated with aircraft status.

The Sense Remote Objects function
This function pertains to the sensing of exterior objects, such as other aircraft or the terrain. A radar system is the primary equipment to perform this function. The received information is provided to the navigation system.

The Provide Power function
All aircraft subsystems require power to operate. The most common power type is electrical, both alternating current (AC) and direct current (DC). In addition, the aircraft uses battery power for backup as well as the two other devices described below. In addition to electrical power, the aircraft uses hydraulic and pneumatic power. But, as Chapter 7 will show, these are only solutions. Therefore, any one of the three functions may not exist if it is not needed. The aircraft could operate on only one power source, if necessary. The optimum power source is the subject of trade-off.

Generators connected to the engines normally provide electrical power, In addition, devices called auxiliary power units (APUs) provide power in emergencies and when the engines are idle. Also for emergencies devices called ram air turbines (RATs) provide hydraulic power in emergencies. These devices were active in the Miracle on the Hudson incident described in Chapter 16.

Other potential functions

The functions described above and in Figures 3.5 and 3.6 are the basic functions required to enable the aircraft to perform its operational mission. Other functions not described above have been suggested. Some systemists, people interested in studying and analyzing systems, argue that these functions are not valid functions since, in their opinion, they do not meet the criterion of a function which is to describe the exchange of energy between systems. They are included here both for completeness and because the reader may find them useful for identifying requirements and performing the other important activities associated with functions.

The Provide Environmental Control function
The primary environmental subfunctions the aircraft has to provide with respect to the passengers and the crew are: Provide Air Conditioning, Provide Pressurization, and Provide Oxygen. Provide Air Conditioning can be subdivided into Control Temperature, Control Humidity, and Control Air Quality. As we will see in Section 9.1, many subsystems will contribute to environmental control, in addition to the ECS itself.

In addition to controlling the environment of the passengers and crew, the aircraft must control its own environment. The primary subfunctions in this category are: Provide Ice Protection and Provide Rain Protection. Ice Protection includes both de-icing and anti-icing.

The primary input to this function is power. This can be electrical power, hydraulic power, or pneumatic power. The output is the energy to condition the aircraft, in the cabin and other parts as described above.

The Provide Passenger and Crew Accommodations function
This function includes a number of subfunctions, the primary ones of which include: Provide Passenger and Crew Space, Provide Seating, Provide Storage, Provide Lavatory Accommodations, Provide Galley Accommodations, and Provide Entertainment. The interiors segment described in Section 9.4 has the primary responsibility for meeting the requirements associated with these functions. The airframe segment, of course, provides the space for passengers and crew. There are other functions related to passenger life support and comfort. These include, for example, the pressurization, temperature control, and oxygen provided by the Provide Environmental Control function discussed in Section 3.3.

Emergency passenger provisions are established by the Perform Emergency Operations function described in Section 3.3, one of the situational functions.

The Provide Cargo Capability function
The primary subfunction associated with this function is Provide Cargo Space. Also included within this function is the Provide Cargo Loading subfunction. This subfunction includes the ability of the aircraft to allow for cargo ingress and egress. In addition, it can allow for autonomous cargo loading and handling systems described in Section 2.4.

Other functions associated with cargo can be found in other sections. For example, the Provide Environmental Control function described above provides for the environmental protection of the cargo. The Maintain Structural Integrity function described in Section 3.3 assures that the aircraft can sustain the cargo loads.

The Maintain Structural Integrity function
The two main subfunctions of the Maintain Structural Integrity function are Sustain Loads and Maintain Pressure, both of which are the primary responsibility of the airframe segment described in Section 9.8.

It is a common misperception that the aircraft structures do not have performance requirements since they "don't do anything." On the contrary, the ability to sustain loads is as valid a function as any other function. The input loads the airframe must satisfy come from a variety of sources: aerodynamic loads, inertial loads, and pressure loads.

This example illustrates the intuitive principle that all components should have at least one function. Although it may seem that the airframe segment described in Section 9.8 has the primary responsibility to fulfill this function, all segments also have to maintain structural integrity.

Structures may have other subfunctions, such as Allow Ingress and Allow Egress. These subfunctions apply to all openings and doors.

It is generally not necessary to examine all flight phases to determine the critical design conditions for structures since the Federal Aviation Regulations (FARs) will provide these. Design conditions will include crash loads, for example.

3.4   Functional Aspects of Safety

A key aspect of aircraft functional analysis is that it is used as the primary tool for identifying and classifying potential safety hazards. This analysis is a critical part of the design for certification as discussed in Section 10.2.

3.5   The Cluster Model

A methodology that provides more insight into a system is called cluster analysis. This methodology was first proposed by Lano (1979) and described by Hitchins (1993, pp. 135–147) and (2003, pp. 143–148, 449–457). The methodology has been incorporated into a software tool called CADRAT by Campbell (2013), which was envisioned by Hitchins.

The essence of the cluster model is that it optimizes the relationships between the elements of a system and presents clusters of entities with *functional affinity*. It is not necessary to know the identities of the elements of a system, for example, the subsystems or components. It is only necessary to know the relationships. The inputs and outputs of the functions in Figure 3.6 constitute the relationships. There is a relationship between *fuel* and *range* for example.

Figure 3.7 shows the output of the cluster model for the top-level entities of a commercial aircraft. These entities were extracted from the IDEF0 model shown in Figure 3.7. So what does the cluster model show? This figure shows that a commercial aircraft is a highly coupled system. If the system were uncoupled, the clusters would represent recognizable subsystems, such as the propulsion subsystem or the avionics subsystem. But as Figure 3.7 shows, the subsystems are apparent at the nodes of the matrix. But most importantly, this figure shows two dominant clusters that can be characterized in the following way:

The first and smaller cluster (from left to right) we call the energy cluster because it contains parameter related to power and thrust and to related functions such as aircraft movement and aircraft environment. This cluster shows that there is functional affinity among those parameters and to parameters they are connected to.

The second and larger cluster is dominated by those entities that pertain to the navigation and control of the aircraft and to related parameters such as information. Hence, there is functional affinity among these parameters as well.

It also shows that aircraft movement is related to parameters in both clusters. Since aircraft movement includes both forward movement and rotational (pitch, yaw, and roll) movement, this entity requires the combined involvement of

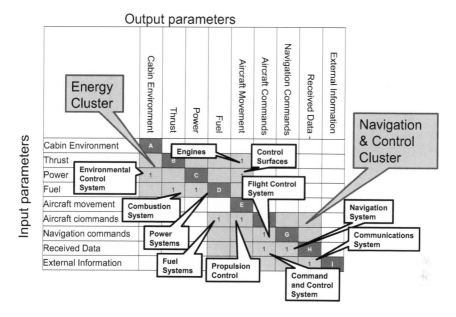

Figure 3.7   Cluster model

many functions. This fact illustrates a certain degree of *loose coupling* in the aircraft system.

As noted above, this model shows that the aircraft is highly coupled, that is, that single entities play a role in many different functions. Two entities that stand out are power, thrust, and aircraft movement. This fact has relevance in the safety context; that is, the loss or malfunction of a single entity could mean the loss of several critical functions. There are many ways to mitigate this risk, many of which are used today. Two of them are as follows:

Functional redundancy
As pointed out in Chapter 16, *functional redundancy* means that a function is performed using two or more physically different means. The use of an APU as a *functionally redundant* source of electrical power is an example.

Modularization
Called localized capacity in Chapter 16, *modularization* means that two or more independent means are used to perform a function. If one means fails, the other will operate independently. Functionally independent engines are an example.

In short, cluster analysis is a useful tool during the Advanced Design phase of development. It will reveal the shortcomings and strengths of any given design and will provide an understanding of the interrelationships among the aircraft parameters.

3.6   The Swim Lane Model

Another commonly used model is the swim lane model. Figure 3.8 shows an example of this model. The advantage of this model is that it shows a clear time dependency among the various elements of the system.

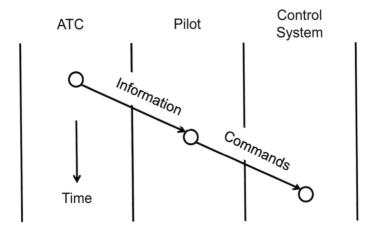

Figure 3.8   The swim lane model

4

Requirements and Needs

The sources for requirements for the development of a new, derivative, or change-based aircraft design fall into two primary categories: regulatory and economic. Regulatory requirements are those which pertain to the safety of the aircraft and its occupants and are established by the Federal Aviation Administration (FAA) and documented in Federal Aviation Regulations (FARs). Economic requirements are those driven by the airline customer and pertain to the cost of purchasing, operating, maintaining, and servicing the aircraft. The SE process accommodates both categories of requirements.

Economic requirements, for the most part, emanate from customer needs, those desires to have an aircraft that economically profitable.

4.1 Requirements Definition

A requirement is a statement of required performance or design constraint to which a product must conform. One principle agreed on by most systems engineers is that a basic quality of a requirement is that it must be verifiable. The requirement is applied to the people, products, and processes, not to the engineer or the environment.

4.2 Requirements Types

Some sources, such as IEEE 1233 (1996), define as many as 25 requirements types. Many systems engineers use only three categories: functional requirements, performance requirements, and constraints. This method of categorizing requirements, used in this book, is simple and logical. The distinction between performance requirements and constraints is only important to the requirements *developer*. It helps him or her strive for the goal of requirements completeness. To the requirements *implementer* this distinction has no bearing on the task of converting the requirements into a design; he or she only has to make sure the design meets the requirement regardless of the requirement category.

Functional requirements

Functional requirements are pretty simple; they are simply the functions as described in Chapter 3 but without any quantitative value for the function. When the quantitative value is added, the performance requirements are born, as described in the following section. Of all requirement types human factors requirements often prove to be of this type.

Performance requirements

A performance requirement is a measure of the extent to which a system performs a function. MIL-STD-961D (1995), for example, encompasses performance requirements within a category called *entity capability*. Performance requirements are a critical part of the certification process discussed in Section 4.9, Item 5. A basic concept is that a*ll performance requirements are traceable to functions.* Ideally, this concept applies even to humans within the system even though satisfying the human requirements is the most challenging of all as discussed in Section 5.5. This challenge is limited, of course, to those functions which apply directly to the satisfactory operation of the entire system. Even so, developing requirements for such critical functions as Maintain Vigilance consumes the energies of human factors specialists.

In the traditional non-SE environment, performance requirements are the most likely requirements to be neglected. A basic concept is that every aircraft element must have at least one performance requirement as described in Section 3.3. This concept is only logical because if it were not true, then elements would have no functions to perform and therefore no reason for existence.

Another concept is the traceability of performance requirements to higher-level requirements. This concept is a logical extension of the aircraft hierarchy in which the function of every element is to support the mission of the entire system. While it is not always practical to create complete top-to-bottom traceability of requirements, the concept assures that the origin of all requirements is known. It also assures that requirements get flowed down to lower levels of the aircraft hierarchy. Standard specification formats, such as the example shown in Appendix 2, often provide for the recording of requirements traceability.

Performance requirements in an aircraft context

We saw before in Chapter 3 that the functional analysis creates a complete functional description of the aircraft, its subsystems, and its components. However, the process reveals nothing about the hardware descriptions. The allocation process creates a performance requirement for each function. (This is only one of three meanings of the word *allocate* (see Glossary).) At the top level, performance requirements will not exist for all functions. However, at a critical level, all functions will result in performance requirements.

At the top level, typical functions, such as Perform Transport Mission, will result in performance requirements, such as, "The aircraft shall be capable of transporting 250 passengers and 20,000 lb of cargo at least 7,000 nmi at a cruise Mach number of at least 0.8." This requirement is *top level*: that is, it does not depend on any design solutions. Requirements which do depend on design solutions are discussed below in Section 4.6. Other performance requirements, such as "Provide 11,000 lb of thrust," are derived from design solutions at the higher level of the system architecture.

We saw the Control Temperature function which was subordinate to the Provide Environmental Control function discussed in Section 3.3. The Control Temperature function is also top level since it does not depend on design solutions. It is necessary to capture all top-level performance requirements before proceeding to the derived requirements.

Performance requirements and interfaces

As we will see in Section 6.1, there is a strong link between interfaces and performance requirements. In brief, for every interface there are at least two performance requirements: For the delivering side of the interface, there will be a performance requirement to create and deliver the quantity being delivered, for example, electrical power. For the receiving side, there will be the requirement to *use* the received function, also, electrical power. For example, a fan will be required to create cooling using the received power.

1. *Emitted quantities*   Certain quantities are emitted from many of the aircraft components. These include noise, radiation, heat, and other quantities. These emitted quantities are often confused with the environments as discussed in Section 5.6. However, there does exist a performance requirement on each component to *limit* the amount of each of these quantities the component may emit. These limiting performance requirements are just as valid and verifiable as any other performance requirements.

   Although these emitted quantities are limited in their value, they do become the environment *for other components* in the immediate vicinity of the emitting component. The most striking example is electromagnetic interference (EMI). Electrical cables are limited in the EMI they may produce. Nevertheless, EMI is a real environment for electronic components. The system is protected from EMI in three ways: by limiting the output of EMI (a performance limit), by controlling the spacing between the cables and the electronic components, and by providing protective shielding for the electronic components.

2. *Delivered quantities and operations on incoming quantities*   In general, the emitted quantities discussed above are generally *undesirable* quantities. However, many other quantities may be intentionally delivered from one component to another. These include electrical current, hydraulic fluid, and pneumatic air. These are *desirable* quantities. Thus, when an interface calls for an electrical current to be delivered from component A (an electrical cable) to component B (a fan), then two performance requirements are implied. First, it is a performance requirement for the cable to deliver the current to the fan. Secondly, it is a performance requirement for the fan to *use* the current to run. Both performance requirements are listed in the appropriate specifications.

Constraints and specialty requirements

Chapter 5 provides a more complete description of the concept of constraints. Suffice it to say that a constraint is any non-performance requirement; that is, any requirement that cannot be traced to a function. That is, it is a *global* requirement. Constraints can include mass properties, dimensions, environments, and design standards. MIL-STD-961D (1995) limits the use of the term *constraints* to such requirements as design standards. Most constraints are established within key engineering specialties, such as human factors, safety, production, and maintainability. The requirement for resilience as described in Chapter 16 also qualifies as a constraint.

In short, constraints are so numerous they often become the design drivers in spite of the fact that much emphasis in SE is on the development of functions, functional requirements, and the resulting performance requirements.

The specialty requirements pose a special challenge to the systems engineer. For the most part they are qualitative. The role of the systems engineer is to convert these qualitative requirements into verifiable requirements in accordance with Chapter 11. Table 4.1 provides some examples of this conversion:

Table 4.1   SE treatment of qualitative requirements

Specialty	Qualitative Requirement	Verifiable Requirement
Human Factors	The item shall be easy to reach.	The item shall be no more than three feet from the seat.
Maintainability	Provide access for replacement and repair of the item.	Provide a minimum of four inches for access on all sides of the item.

Design requirements

Design requirements are the design characteristics which are the product of the synthesis process as described in Section 7.7. They are the direct result of the performance requirements and constraints discussed above. Design requirements are the attributes of the item needed to meet the performance requirements and constraints. These could include, for example, physical dimensions or power required.

4.3   Requirements Development

A frequently asked question in SE is: When are all the requirements complete? The answer is that they are complete when the customers (including the regulatory agencies) say they are. SE provides some tools to assure that as many requirements have been captured as humanly possible. Here are a few steps.

The famous Vee model

One of the most often used ways of depicting the requirements development process is with the Vee model shown in Figure 4.1. Although the Vee model is a somewhat simplistic way of illustrating the process, it does bring out some of the important aspects of SE and especially how SE handles requirements. In short, the Vee model is a concise way of depicting the entire SE process. Following are some of those aspects illustrated by the Vee model:

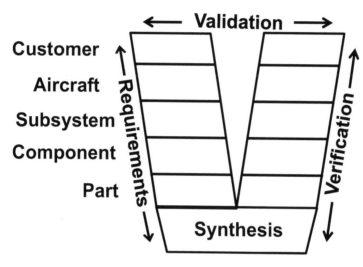

Figure 4.1   The Vee model

The Vee model overview
The left side of the Vee model represents the flow down (and up) of requirements as needed from level to level of the aircraft hierarchy. The right side depicts the flow up (and down) of the verification of the various requirements. Chapter 11 discusses verification. At the top of the Vee model the validation of customer needs is shown from left to right. Validation is also discussed in Chapter 11.

As will be explained below, the Vee model is only part of the picture. It represents the *reductionist* view of requirements development. That is, it only represents those requirements determined from the flow down from higher levels of the aircraft hierarchy. The broader view is called the *holistic* view which is discussed below. The holistic view accounts for requirements from other sources than flow down.

The aircraft abstract hierarchy
Chapter 2 explained how the aircraft system could be depicted in terms of an abstract hierarchy as in Figure 2.1 with the aircraft at the top and subsystems and so forth at lower levels. The Vee model shows that requirements are flowed down

and allocated through the various levels of this hierarchy and that the verification tasks are flowed up through the same levels.

The time axis
The horizontal axis is time. Hence, the Vee model reflects the phased development inherent in integrated product development (IPD) discussed in Chapter 12.

Customer needs
Customer needs (also called customer requirements) are shown at the top of the Vee. As explained later in this chapter, the customer needs need to be converted into top-level aircraft (product) requirements before the flow down process can begin. It is important to distinguish customer needs from system, or product, requirements. Needs are often qualitative. It is not even known whether the needs can be achieved. It will be shown later in this chapter that a basic quality of a requirement in achievability. This is not so for needs; needs are not even verifiable since they are not product requirements. Working with the customer, the developer can conduct trade studies and other analyses to convert the needs into requirements. Section 7.4 will show that one way to convert needs into requirements is with quality function deployment (QFD) analysis.

Requirements flow down
Starting with the top-level aircraft product requirements (range, durability, and so forth), the Vee model shows the flow down all the way to the parts by allocation and requirements derivation as explained later in this chapter.

Synthesis
As explained in Chapter 7, synthesis occurs when the entire aircraft is integrated and assembled into a unified system. This occurs after all requirements have been defined.

Requirements flow up
In addition to requirements flow down as explained above, the Vee model also shows that requirements can flow up. This happens when it is desirable to use a commercial off-the-shelf product (COTS) to perform a certain function perhaps at the component level, such as a pump or a fan. When this happens, it is necessary to flow the requirements up to the aircraft level to confirm that it is not in conflict with any top-level requirements.

Verification flow up
Since verification must be performed at the same level as the requirement, the verification steps will be performed from bottom up as shown in the Vee model. For example, there will be verification tests (as required) for the parts, then components, and so forth. Most aircraft developers have laboratories in which complete subsystems can be tested, such as hydraulic systems, control systems, and so forth. Verification at the top level usually consists of flight tests.

Validation

As explained in Chapter 11, validation of customer needs must occur at the customer level. This can consist of flight demonstration tests (by the customer) or by buy-off of the lower-level tests. This is called system validation and should not be confused with requirements validation, which according to the FAA (2012, p. 54) is the assurance that the requirements are correct.

Reductionism vs. Holism

At the risk of introducing some somewhat esoteric terms, it is important to understand the difference between *reductionism* and *holism*. Many books have been written, for example Checkland (1999, p. 77), explaining why reductionism is undesirable and holism is desirable. In short, reductionism is vertical thinking, while holism is thinking about the whole, including the horizontal relationships between elements.

Hence, using the Vee model, as described above, alone without any consideration for the relationships across elements is reductionist and therefore incomplete. For example, you have a system A with two subsystems A_1 and A_2. Using the Vee model you could determine the requirements for A_1 and A_2, but you would know nothing about the relationship between A_1 and A_2.

Among SE sources, the one that addresses the holistic approach is Stevens et al. (1998, pp. 6, 206, 344). According to Stevens et al., the additional relationships that should be addressed include:

- the system architecture;
- the various disciplines involved (electrical, mechanical, and so forth);
- keeping requirements (including interface requirements), design, plans, and risks consistent;
- the various system development processes (engineering, integration, and so forth);
- requirements, costs, and timescales.

The *FAA Systems Engineering Manual* (2014) also notes the importance of including the requirements for the operational phase as part of the holistic considerations.

So what do you do if you find that the reductionist and holistic requirements are in conflict? The answer is that you conduct a trade study. Section 4.8 discusses the principles for how to conduct requirements trade-offs.

Mission statement

The mission statement is the first step in the requirements development process. The mission statement is a simple statement of the purpose of the system being developed, the environment in which it will operate, and any special operational considerations which may be important. The value of the mission statement is to establish an understanding between the customer and the technical community about what the system is for. The mission statement will be the primary exhibit at the system requirements review (SRR) described in Chapter 12.

Mission statements apply at any level of the aircraft hierarchy. For example, at the aircraft level, the following simplified mission statement might be adequate:

> The purpose of the aircraft is to carry 250 passengers and 20,000 lb of cargo a total distance of 7000 nmi at a Mach number of 0.8. The aircraft will operate primarily in sandy desert climates.

At a subsystem level:

> The purpose of the subsystem is to protect the environmental control system from damage due to particulates in a sandy desert environment. The subsystem will remove at least 90 percent of the particulates from the air it receives.

Requirements allocation from functions

The allocation of functions to performance requirements is the basic completeness technique of SE as described in Section 4.2. This meaning of *allocation* is the second of three definitions (see Glossary). The principle of allocation of functions to performance requirements is part of the certification process.

4.4   Requirements Sources

A basic practice of SE is to identify, justify, and record the source of all requirements. This process is called requirements validation. The principle of requirements validation is reinforced by the certification process which requires that all requirements be validated and recorded as shown in Table 10.1 as part of the certification data package. Furthermore, certification requires that a validation plan be submitted and that data be submitted to substantiate the validation. Section 4.10 explains requirements validation in more detail.

External requirements

External requirements are those which emanate either from the airline customer, regulatory agencies, such as the FAA or Joint Aviation Authorities (JAA), or industrial organizations.

Customer requirements
Unlike for the automobile industry, the requirements for an individual airline customer may result in a product which is highly focused on that customer. The process of capturing those requirements, negotiating with the customer, and incorporating the requirements into an aircraft can be a highly complex process.

Customer requirements are sometimes, and more accurately, called customer needs as described in the Vee model of Figure 4.1. This is because customer requirements are not necessarily achievable on first sight. The developer needs to work with the customer and develop a set of mutually agreeable requirements that can be flowed down to the aircraft level as shown in Figure 4.1. Requirements at the aircraft level are product requirements and must meet the criteria of *achievable* and *verifiable*.

The important aspect of capturing customer requirements is to be able to ascertain what the customer wants the product (either the aircraft or a subsystem) to do, that is, to understand the functionality of the need.

One common mistake is to accept buyer furnished equipment (BFE) unquestioningly without examining its requirements. There are some key questions which need to be answered: Does the equipment do what the customer expects it to? Will the equipment function in the environment of the host aircraft? Will the equipment interface with the aircraft in an acceptable manner? Will the equipment meet the safety requirements which have been imposed on the rest of the aircraft? All of these questions should be answered before the equipment is incorporated into the aircraft. The results should be briefed to the customer at the SRR as described in Section 12.4.

One method for capturing customer requirements is QFD. QFD is a structured methodology for conducting dialogue with a customer, prioritizing the customer's needs, identifying solutions to those needs, and scoring those solutions. QFD has been used successfully in the automobile industry and shows great promise for the aircraft industry.

Another type of customer requirements is *assumed* requirements. That is, during requirements development some requirements have to be assumed. For example, the number of cubic feet of storage area in the cabin is a requirement designated as assumed. The assumed value of the requirement is presented to the customer at the SRR for concurrence. If there is no concurrence, the requirement can be changed.

Regulatory requirements

Traditionally the regulatory requirements have been passed to the aircraft manufacturers through the FARs and have been verified in a certification plan submitted by the manufacturer. A major part of the certification plan is a Functional Hazard Assessment (FHA) which identifies hazard categories for identified components so that the proper designs and redundancies can be implemented.

The FAA, in cooperation with the Society of Automotive Engineers (SAE), has taken a major step towards incorporating the principles of SE into the certification process with the publication of the SAE ARP 4754A (2010). The guidelines of this document are not mandatory but recommended processes. It recommends that a thorough SE functional analysis be conducted for all levels of aircraft systems and that the hazard category be established for each function. It also recommends that combinations of functions be identified for potential hazards. It is expected that this process will result in increased aircraft safety. Ultimately, these recommended practices may be incorporated into an FAA advisory circular (AC) and eventually into a FAR.

Industry standards
Many standards are developed by industrial organizations, such as the SAE, the Institute of Electrical and Electronics Engineering (IEEE), and RTCA, Inc. The requirements validation should cite these sources. Examples of particular interest to the commercial aircraft industry include the SAE ARP 4754A and RTCA/DO-178B.

Internal requirements

Many requirements originate from internal sources rather than from the airline customer. These requirements have many objectives: for example, to reduce cost, reduce weight, improve reliability, or simply to fix an item that does not work properly.

It is tempting to treat internal requirements less rigorously than, say, customer or regulatory requirements. Suffice it to say that internal requirements demand at least as much rigor as any external requirements. Just as much is at stake, for example, cost, safety, performance, reliability, to name a few of the more common ones. Before the redesign or fixing of a part or a subsystem, it should be established what the item was intended to accomplish, what environments it operated in, what other constraints were important, and what it had to interface with, both functionally and physically.

One innovation of recent years is the *quality circle*, a part of total quality management (TQM). In this process a team of employees attempts to resolve product or process problems in a group environment. This process has been regarded as highly successful. The combination of TQM with the SE rigor of defining requirements can result in value-added improvements. This approach can be applied to any of the requirements areas listed below.

Manufacturing initiated requirements
Many requirements originate in the manufacturing department of the aircraft company after the aircraft has been designed. The purpose of these requirements is usually to make the product easier to manufacture. Like most internal requirements, if the product had included manufacturing inputs during the initial aircraft development, these requirements would be unnecessary. However, they exist and should be included as a valid requirements category.

Product improvement
Design organizations will often initiate design improvements internally to improve a product. The purpose of these improvements may be to improve either weight or cost. Like manufacturing initiated requirements, these are after-the-fact requirements to correct oversights in the original design.

Design for manufacturing and assembly (DFMA)

The basic purpose of DFMA is to reduce the recurring cost of an assembly. Other goals may be weight savings or improved reliability. These goals are normally accomplished by reducing the number of parts on a certain part of an aircraft. Other design improvements may accomplish the goals. When an assembly is identified as a potential area for improvement, a study is initiated to determine the cost savings, for example, of the redesigned assembly. If the non-recurring cost is returned in a given number of aircraft, say 20, then the project will be initiated.

Before the DFMA project can be initiated, both performance requirements and constraints for the original assembly should be determined and documented. The following questions are asked: What function did the original assembly have to perform? What were its performance requirements? What were its constraints? What environment did it have to withstand? What interfaces did it have with surrounding components? If the original requirements were not well documented, or perhaps not documented at all (after all, it was probably designed before SE was fully introduced into the organization), then the second question should be rephrased: What were its performance *capabilities*?

Product support initiated

The purpose of this type of change initiative is to improve the design of the aircraft so that it can be more easily maintained. For example, the structure surrounding a line replaceable unit (LRU) may be modified to allow a person to replace it more easily.

Technology focus initiated

This type of change is normally used to correct an operational deficiency. For example, an LRU may be replaced to provide a more reliable model.

4.5 Requirements Allocation to System Elements

This is the last major step in the requirements subprocess of the SE process. Here is where we decide what elements the requirements pertain to. At this point the shape of the element may not be known. It is only a *bucket* on the spec tree of Figure 2.1. Requirements may be allocated to multiple elements. For example, noise requirements are allocated to many elements. The allocation of requirements to elements is a basic requirement of certification.

4.6 Derived Requirements

Derived requirements are those requirements which depend on some feature of the solution to determine their values. For example, the value of engine thrust is a derived requirement determined from extensive trade-offs in the conceptual design process as discussed in Section 8.2. Most, but not all, subsystem requirements are, indeed, derived. Hence, derived requirements cannot be determined at the outset of the program. They are established gradually as the program progresses.

Another way to look at derived requirements is through the principle of requirements *harmonics*. Figure 4.2 illustrates this principle. In this figure the top-level requirement results in a solution, which, in turn, results in anther requirement (the derived requirement) which results in another solution. These alternating requirements and solutions are called requirements harmonics.

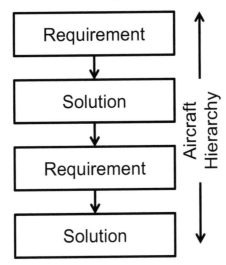

Figure 4.2   Requirements harmonics

Some sources, for example ARP 4754A (2010, p. 11), point out that derived requirements are not directly traceable to higher-level requirements since they may be traceable to design decisions. The key word here is *directly*; hence, since the design decision itself is traceable to higher-level requirements, the derived requirement will be indirectly traceable to the higher-level decisions. In some cases the derived requirement may be mathematically determined from higher-level requirements. For example, the requirement for the energy absorbed by a brake pad can be determined from the laws of physics only knowing the stopping distance, the weight of the aircraft, and the landing speed. The only design decision involved in this calculation is the number of brake pads. Another derived requirement is the amount of current for which a fuel line coupling must be spark free due to a lightning strike. This derived requirement depends on the surface area of the coupling.

4.7   The Principle of Top-Down Allocation

This section will show how to allocate requirements from higher levels of the aircraft hierarchy as discussed in Section 2.3 to lower levels. For example, the analyst allocates requirements from the aircraft level to the subsystem level or from the subsystem level to the component level. This type of allocation corresponds to the third definition in the Glossary, namely, the breakdown of a top-level requirement into its subordinate components.

The principle of allocation assumes some degree of knowledge of the solution, that is, the design of the aircraft. Therefore, allocation will occur only after the appropriate level of aircraft design has been defined. For example, at the system design review (SDR) as discussed in Section 12.4 the aircraft architecture will have been defined, such as the one shown in Figure 2.1. Furthermore, at the SDR, an initial concept will have been developed. From this architecture and the initial concept we will be able to estimate the weight of each subsystem based on past experience. This estimate then becomes the *initial* weight allocation for each subsystem. We say *initial* because during the trade-offs discussed in Section 7.2 leading up to Preliminary Design Review (PDR) as described in Section 12.4 these allocations may, and probably will, change as a result of requirements trade-offs as discussed in Section 4.8.

Other allocated requirements, such as loads, shock, vibration, and noise, may require a greater degree of concept formulation to perform allocation. This requirement results from the fact that the allocation of these parameters requires the definition of the structure of the aircraft. In any event all allocation should be performed before PDR since all requirements are defined by PDR.

Requirements allocation is more than just a method of determining requirements at lower levels. It is also a method of program control. Some parameters, for example, weight and dispatch reliability, among others, are primary candidates for technical performance measures (TPMs) discussed in Section 12.7.

Weight

In commercial aircraft, weight is one of the most important design parameters. One hundred pounds of weight can have a significant effect on the operating costs and on the range of the aircraft. That is why it is one of the primary TPMs, as described above.

Weight allocation from the aircraft level normally starts with the Manufacturer's Empty Weight (MEW) discussed in Section 5.2. MEW is the weight of all the aircraft structure and components, without any fuel, crew, passengers, or cargo.

Weight allocation is mathematically simple since weight is additive. As described above, the allocated weights result from an initial concept formulation. These allocations can then be adjusted as the final design takes shape. But, in any event, the objective is to keep the total weight of the aircraft (the *sum* of the allocated weights) below the initial constrained value.

Non-recurring (development) cost

Although aircraft manufacturers eventually amortize development costs over the unit costs of aircraft, they, nevertheless, set limits on development costs which they allocate to the aircraft components.

Recurring (unit) cost

We will see in Section 8.6 that recurring cost is a major component of the direct operating cost (DOC), which is a major design driver. Hence, allocating recurring cost to the aircraft elements is of primary importance in controlling the total aircraft cost. Successful cost control of each element would eliminate the need for cost reduction activities, such as design for manufacture and assembly (DFMA), as discussed in Section 4.4.

Direct Operating Costs (DOC)

Although DOC itself is not directly allocated to the aircraft elements, its components are described in Section 8.6. In addition, many derived requirements, such as weight, aerodynamic parameters, and fuel consumption, have direct impact on DOC. This dependence allows the management of DOC through these parameters.

Dispatch reliability

Dispatch reliability is one of the main drivers of airline indirect costs discussed in Section 8.6. Hence, manufacturers and airlines like to keep it as high as possible, normally in the range of 0.99. Allocation of dispatch reliability is not as straightforward as additive parameters, such as weight. Reliabilities must be multiplied to determine total reliability. Appendix 1 provides a simplified explanation of how to allocate reliability.

Maximum Allowable Probability (MAP) of failure

Like dispatch reliability discussed in Section 4.7, the maximum allowable probability (MAP) of failure can be allocated to all aircraft elements. This allocation is an integral part of the safety analysis discussed in Section 10.2.

Internal noise (sound levels)

The internal noise is the noise that each aircraft element emits, and not the noise environment it must withstand discussed in Section 5.9. Noise allocation is dependent on two factors: frequency and the location at which it is measured. Frequency is straightforward: the noise should be allocated at each frequency band of concern. Location is somewhat more complex. For the simple case of a sound emitter in the cabin, such as an air conditioning duct, the noise the duct can emit is limited by the location of the passengers with respect to the duct. When the noise allocation is made, specific cabin locations will have to be specified.

For noise emitters outside the cabin, the noise allocation will depend to a great extent on the insulation capabilities of the cabin walls, that is, on the design solution of the aircraft itself. Hence, this allocation cannot be made until the cabin wall concept is formulated. Hence, it is a derived requirement.

External noise

External noise, also frequency dependent, and primarily from the engines, is primarily limited by regulatory requirements pertaining to airport noise limitations. Allocation of noise to other components will be limited by their effect on ground crews.

Electrical loads

Electrical load analysis will determine the loads delivered to various components of the aircraft. This analysis can be used either to resize the generator (and hence place a penalty on the engine output) or to limit the electrical power delivered to various components.

Air distribution

For the most part, air distribution is the responsibility of a single subsystem, the environmental control subsystem (ECS). Nevertheless, the ECS needs to allocate the air distribution to the various compartments and areas: passenger cabin, flight deck, lavatories, galley, cargo, and other special areas.

Fuel consumption

Fuel consumption can be directly allocated to four factors: propulsive thrust, electrical power, hydraulic power, and air bleed for pneumatics. The degree to which each one of these consumes fuel is strongly dependent on the demands for each of these quantities. Minimum fuel consumption results from trading off these power drains. As discussed above, fuel consumption has a direct effect on DOC.

Emissions

Toxic emissions are a major concern in the design of modern aircraft. These are primarily the responsibility of the power plant and the APU.

Maintenance cost

Total maintenance cost is a component of DOC discussed in Section 8.6. Maintenance cost is a prime candidate for allocation to the various aircraft elements. The cost allocation will depend to a great extent on the element type, for example, electrical, electronic, mechanical, or structural.

Loads, shock, vibration

More than most other allocated parameters, loads, shock, and vibration are strongly dependent on the aircraft design. Hence, analyses will determine how these parameters are transmitted through the structure tc various parts of the aircraft. These analyses will determine how they affect different components. This allocation can then be used either to limit the source loads or to design the structure to limit the transmission of the loads.

Durability

A driving requirement in the commercial aircraft domain is durability. The question is how many take-offs and landings can the aircraft make before cracks begin to appear in the structure? This requirement is particularly important in the regional jet market. The reason this requirement is important is because regional jets make far more take-offs and landings than larger jets. Another reason this is important is because metal fatigue is notoriously difficult to predict. For composite structures the issue is delamination, which is also difficult to predict. Either way, the only path to having a durable aircraft is to use the best data and the best prediction methods available and then add a good safety margin.

The allocation method for durability is simple: it is that every component on the aircraft has to meet the same requirement as the whole aircraft. Of course, structural components will receive the most attention for this requirement.

4.8   Requirements Trade-Offs

Chapter 7 discusses design trade-offs; this is the most common understanding of trade-off. Requirements trade-offs are less familiar. For example, at the aircraft level, there are trade-offs to minimize the take-off weight (W_{to}) and the DOC. Another derived requirement, the drag coefficient of the aircraft, CD, results from minimizing these parameters. This process is called a requirements trade-off.

A second aspect of requirements trade-offs pertains to allocated requirements, as discussed above. We saw how to establish certain requirements by top-down allocation that is dividing a top-level parameter, such as dispatch reliability, into parts and allocating them to different elements as described in Section 4.7. The first cut at this allocation is just a guess based on estimates of past reliabilities of those elements. In subsequent passes you can refine those estimates by trade-offs. This process answers the question: What allocation of dispatch reliabilities will result in the highest total dispatch reliability? Varying the allocation may result in a higher level of dispatch reliability. This procedure is another requirements trade-off.

Requirements trade-offs and examples

Amazingly SE literature provides very little guidance and very few examples of how to perform requirements trade-offs Figure 4.3 shows in a conceptual way the reason that vertically flowed down requirements may be in conflict with requirements from other sources. In this diagram requirements are shown flowing down from a top-level system called System A to two lower-level subsystems called Subsystem A_1 and Subsystem A_2.

As was explained above, requirements can be flowed down from A to A_1 and A_2 by various means including allocation, derivation, and straight flow down. In addition, both A_1 and A_2 may be the recipients of requirements from other sources including interfaces, production, operations, and differing requirements from different stakeholders. Some requirements conflict simply because they were incorrectly allocated. Since these requirements are not flowed down, this figure shows them as horizontal entries. Hence, the flowed down requirements, shown as vertical and the horizontal requirements, may be in conflict since they come from different sources; there is a need for a trade-off between them. The nature of the trade-offs will be determined by the situation and the nature of the requirements.

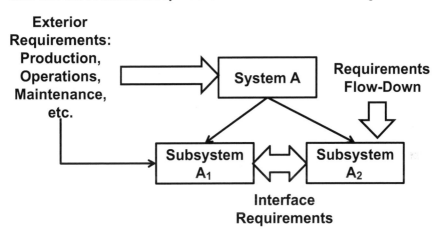

Figure 4.3   Sources of conflicting requirements

Linear addition
In some cases requirements from different sources may be added together. For example, the energy the brake pads need to absorb can be derived by a simple calculation which depends on primarily the weight of the aircraft, the landing speed, and the length required for stopping. Let's call that energy E_1. However, additional energy may be required for stopping on wet runways. Let's call that energy E_2. In addition, the customer may want to carry an additional amount of cargo. This will result in an additional amount of energy E_3. Hence, it may be

assumed that these energy requirements may be added together resulting in a total energy $E_1 + E_2 + E_3$. This linear assumption may not always be valid, so the analyst will need to examine it carefully.

Modeling
In other cases the combination of requirements is so complex the only way to combine them is through a computer model of the system. Once again the example selected pertains to an aircraft brake system and the complex phenomenon of friction-induced instabilities as described by Hamzeh et al. (1999).

Requirements weighting
In some cases the correct requirement may be determined by examining how the conflicting requirements would affect the design based on a metric such as cost or weight. This technique would most likely involve a preliminary design based on the conflicting requirements. Such techniques, such as parametric cost analysis (PCA) could be used.

The QFD method described in Section 7.4 uses the weighting method. This method is particularly applicable when customer requirements are in conflict.

Absolute limits
Sometimes requirements are not tradable. Take maintenance requirements for example. The space required to repair or replace a component is limited. If this requirement is in conflict with another requirement, the space requirement will probably prevail. In general, though, when two requirements are in conflict and one of them is more restrictive than the other, the more restrictive one will prevail. Of all requirement types safety will most likely prove to be the most restrictive.

Reallocation
One of the primary methods of flowing down requirements is allocation. Many requirements are allocated, such as maintenance labor-hours, reliability, and weight. One way to resolve a conflict is to reallocate the requirements among the components they were allocated to.

4.9   Requirements Categories for Certification

The categories of requirements data required for certification are in agreement with the traditional SE categories but are broken out as shown below (1–8), as summarized from ARP 4754A (2010). The requirements list is comprehensive, showing that certification is concerned about the complete requirements formulation of an aircraft.

1. *Safety requirements* For certification purposes, safety requirements are identified separately because of their primary importance in the certification process. Safety is discussed more thoroughly in Section 10.2.
2. *Functional requirements* For certification, the term *functional requirements* refers to all requirements except safety requirements. These include customer requirements, operational, performance, installation and physical, maintainability, and interface requirements.
3. *Customer requirements* Customer requirements include both market-driven and specific customer requirements as described in Section 4.4.
4. *Operational requirements* These are the requirements associated with the interface between either the flight crew or the maintenance crew and the aircraft.
5. *Performance requirements* For the purposes of certification, performance requirements are in agreement with the SE definition discussed in Section 4.2.
1. *Physical and installation requirements* Many of the physical and installation aspects, for example, mounting provisions, will also appear as interface considerations, as shown below in Item 8. Others, for example, access, will be part of maintainability requirements as described in the following section.
2. *Maintainability requirements* Maintainability in a certification context is compatible with the SE view of maintainability discussed in Section 5.7. Although certification focuses on safety related maintainability, this category of requirements also includes economic related maintainability discussed in Section 8.6.
3. *Interface requirements* This category concerns the functional and physical interfaces which are part of the standard SE methodology as described in Sections 6.1 and 6.2.

4.10 Requirement Validation

Finally, before we enter these requirements into a specification, we have to ask ourselves: Are these valid requirements? This process is called requirements validation, as distinct from system validation discussed in Chapter 11. There is a standard list of tests for requirements validity. ARP 4754A (2010, p. 54) provides such a list. This list elaborates on that list and provides some guidance of how to achieve some of the subtler aspects of requirements validity.

Verifiable
Above all, requirements must be verifiable by one of the four methods described in Chapter 11, namely, test, demonstration, analysis, or inspection. To meet this criterion the requirement must be physically possible.

Not vague or unambiguous

Vague and ambiguous requirements are some of the more common problems with requirements. For example the requirement, *the [item] shall be as efficient as possible,* is vague. This problem calls for a trade-off to determine exactly how efficient the item should be. The trade-off options in Section 4.8 should provide a solution to vagueness.

Explicitly stated

By *explicit* it is meant that the requirement must state exactly what system or subsystem must meet the requirement and what the system or subsystem must be required to do. The template for a requirements statement is often put this way:

The [system or subsystem] shall [perform the specified action] in the [specified environment].

There are several things to notice about this template:

- First, by *system or subsystem* the requirement may apply to any level of the system architecture.
- By *shall* the requirement is mandatory.
- By *specified action* the requirement performs some specified quantitative goal. This item should also include specific tolerances on the performance of the action.
- By *specified environment* the environment may be specified within the statement, or it may reference an environment specification, such as a specification documenting the environment in various zones of the aircraft.

The use of this template will demonstrate that the requirement is identifiable as a requirement.

Achievable

If the implementation of a requirement requires the rewriting of the laws of physics, then this requirement needs to be rewritten. However, achievability may be limited by more practical aspects, such as the lack of space, time, or technological maturity of a component.

Justified by supporting data

This means that the appropriate analyses have been conducted with correct data. This test may apply to requirements that have mathematically derived from a higher level in the system hierarchy. Requirements with error of fact will not meet this criterion.

Not in conflict

Requirements may be in conflict if they emanate from different sources. For example, as discussed above, the requirements developed from flow down may be in conflict with an interface requirement. The solution to resolving conflicting requirements is though a trade-off as described in Section 4.8.

Necessary

It is not always easy to tell when a requirement is not necessary. The way to tell whether a requirement is necessary or not is to see what deleting it does to the design. If the design does not change when the requirement is deleted, the requirement is not necessary and can be deleted.

Not redundant

Like unnecessary requirements redundant requirements are not always easy to recognize if they are worded differently. However, the test is the same; if a redundant requirement can be deleted without changing the design, it should be deleted also.

How well, not what

For example the statement *a fan shall be installed* is not a valid requirement. It is a statement of a task. It does not say how well the fan should work or how much air it should move.

Understandable by supplier

A common problem is that supplier requirements are often written in terms a supplier cannot understand usually due to the physics of the situation. For example, the requirements may be written in terms of decibels, while the supplier only understands vibration. The requirement should be rewritten in term of vibration.

Valid source

The requirement should come from a valid source. Valid sources include the customer, regulatory agencies, and rigorous derivation from higher-level requirements.

Part of a complete set

The requirement when included in the larger set of requirements should form a complete set of requirements.

Complete and valid safety set

Are all requirements necessary for a safety analysis complete and valid?

Allow for flexibility and expansion

As described in Section 5.13, the requirements for new designs often allow for the design of derivative aircraft as defined in Section 2.2.

4.11 Avoiding Requirement Creep

Having read the previous sections of this chapter, readers may feel overwhelmed. The problem the readers will counter is the sheer number requirements. Every specification will have, for example, at least one performance requirement. On top of this there may be hundreds of unnecessary requirements in the tool. Deciding how to decide may be a more difficult task. Here is a simple scheme for deciding:

Assign three people to the task of deciding which requirements or categories of requirements are in or out of the tool. The following three people are suggested:

The Chief Engineer or Project Manager

More than anyone else, the chief engineer should have a comprehensive understanding of the technical aspects of a project and the specifications associated with the project.

A common practice for the Chief Engineer or Project Manager to do is to prepare a document with a name like Project Description. This is a very preliminary description of the project before any design work has begun. One of the features of this document is to describe the project in terms of the disciplines required. This document should aid all participants on the project, especially technical specialists in determining whether they have any role on the project or not.

The Technical Specialist

Certainly, if the component being procured has any software in it, the judgment of the software engineer will be necessary. The identity of any technical specialty will depend, to a great extent, on the nature of the component. If there are any electrical parts in the component, both the electrical engineer and the EMI specialist will be required. More importantly, if these components are *not* present, then the requirements for them can be omitted. This is what we are trying to accomplish: the elimination of requirements that are not needed and would clutter up our tool and specifications.

If any advanced technologies are involved, such as composite materials, the participation of specialists in those technologies would be essential. These specialists could then indentify associated specialists, such as production engineers, who could identify the production requirements associated with composite materials.

The Systems Engineer

Although systems engineers are normally associated with the *process* of SE, their broad familiarity with the many aspects of the design, across disciplines, make them particularly able to identify the particular requirements that are either necessary or unnecessary.

They are also adept at the methodologies for resolving conflicts between requirements, as described above.

Summary of requirements creep adaptation

As will be seen in Chapter 13 "Adapting the Systems Engineering Process to the Commercial Aircraft Domain," adaptation is all about risk. That is, you want to gain the maximum effectiveness while at the same time minimizing the risk of deleting steps or processes. Requirements screening is one of those processes in which this principle is paramount.

Therefore question that the three requirements screeners, above, have to keep in mind is: can this specialty be ignored at this phase without causing any undue risks? They are the only ones who can make this judgment; however, it is a necessary step.

5

Constraints and Specialty Requirements

The term *constraints* refers to all non-performance requirements, that is, requirements which cannot be determined from functions. Whereas performance requirements establish how well a system or subsystem should perform, constraints define the limits on that performance. As we have noted, definitions differ among systems engineers. For example, MIL-STD-961D (1995) uses the term *constraints* for a limited set of design constraints. The present definition, though, is preferred for clarity and simplicity.

Specialty requirements are those requirements set by the various engineering specialties. These include, but are not necessarily limited to, human factors, reliability, maintainability, safety, environments, mass properties, and software. Because of their importance in the certification process, safety and software are discussed in Chapter 10. Other specialties include tooling, manufacturing, and facilities. Constraints and specialty requirements are discussed together here because many constraints arise from the engineering specialties. However, specialty requirements may be either performance requirements or constraints.

5.1 Regulatory Requirements

There are two primary sources of regulatory requirements: First are the civil aviation agencies, the US Federal Aviation Administration (FAA) and the Joint Aviation Authorities (JAA), the latter representing other, primarily European, countries. Other countries, such as Canada and Russia, also have their own aviation regulatory requirements. The primary interest of all regulatory agencies is the safety of the aircraft, crew, and passengers as discussed in Chapter 10. These agencies regulate the entire aircraft manufacture and operation system, including manufacturers, airlines, airports, and air traffic control.

The second source of regulatory requirements is the environmental requirements established by such agencies as the Occupational Safety and Health Administration (OSHA) and the Food and Drug Administration (FDA). These agencies are responsible for controlling toxic emissions as well as other hazardous materials which might be a threat to the passengers, crew, and the public as well as sanitary construction of airplane water systems and galleys.

Like all requirements, regulatory requirements are subject to verification in accordance with the principles of verification discussed in Section 11.3 of which certification is a major part.

The FAA (2012, p. 4) has provided a list of example regulatory documents. They are as follows:

- FAR 25.1301 "General Requirements for Intended Function." This document pertains to the installation and function of individual items of equipment. It states that each item must:
  - be of a kind and design and appropriate to its function;
  - be labeled as to its identification, function, or operating limitations, or any applicable combination of these factors;
  - be installed according to its limitations specified for that equipment; and
  - function properly when installed.
- FAR 25.1309 "Equipment Systems and Installation." The importance of this document is that it establishes the relationship between the severity of a failure and the probability of its occurrence.
- AC 20-152 "Invokes RTCA DO-254 Design Assurance Guidance for Airborne Electronic Hardware." This is a document providing guidance for the development of airborne electronic hardware.
- AC 20-115B "Invokes RTCA DO-178B Software Guidance." This document is used for guidance to determine if the software will perform reliably in an airborne environment.
- AC 20-174 "Invokes ARP 4754A (2010) Guidelines for Development of Civil Aircraft and Systems." This document is used as guidance for the certification of aircraft systems, meaning subsystems in the context of this book.
- ARP 4761 "Guidelines and Methods for Conducting the Safety Assessment Process on Civil Airborne Systems." This document is used for guidance in determining the compliance of aircraft systems with FAR 25.1309 (described above).

5.2 Mass Properties

The term *mass properties* refers to all parameters pertaining to the aircraft's weight and its distribution, such as weight, center of gravity (c.g.), and moment of inertia.

Weight

Weight is a requirement which generates much attention in the commercial aircraft industry. During preliminary design a weight goal is established for all future work discussed in Section 8.2. The weight of the aircraft is then broken down and allocated to the major subsystems shown in Figure 2.1. In this way, weight becomes a derived performance requirement since the weight is a critical factor in determining the range of the aircraft.

Weight may also be constrained by airport factors, such as the strength of the runway. Many airports limit the weight of the aircraft which may land. This type of weight limit is a true constraint.

Weight can be limited by transportation factors. For example, if a segment of the aircraft, for example, the fuselage or wing must be transported by rail, road, or air from a supplier location, it is often necessary to break the aircraft into segments for transportation as discussed in Section 5.12. The size of each segment may be limited by weight. This weight should be specified in the design of the segment. Military specifications, such as MIL-STD-961D (1995), often have special sections for transportation limits. However, if these limitations, such as weight, are taken into account in the weight limits, this section is not necessary.

The weight of the passengers and cargo is a performance requirement as discussed in Section 4.2. From this requirement and other performance requirements (range, speed, and so on), we calculate the total weight of the aircraft as described in Section 8.2. However, weight as a constraint sets the boundaries on the aircraft performance. As shown above, weight is also a factor in DOC and hence is used to manage DOC.

Weight can be expressed in different ways. Each method constrains the design in a separate way. Following are the principal weight parameters:

Manufacturer's empty weight (MEW)
MEW is the primary weight parameter used for design. It includes all system and subsystem equipment items on the aircraft. It does not include the crew, fuel, cargo, luggage, expendable items, such as food, or crew items. MEW is the basic weight which can be broken down and allocated to the various subsystems.

Maximum take-off weight (MTOW)
MTOW is important from a performance point of view since it is a parameter in determining the take-off distance. It is also the parameter used to assure that the aircraft stays within the runway strength limits, as mentioned above.

Other weights
Other weights, whose significance is obvious by their names, are as follows:

- taxi weight
- landing weight
- zero fuel weight
- maximum jacking weight
- maximum towing weight.

Each type of weight should be specified and managed to control its particular aspect of aircraft design.

Center of gravity (c.g.)

The c.g. of the aircraft is one of two principal parameters which together determine the stability of the aircraft. The other parameter is the center of pressure (c.p.). The distance between the c.p. and the c.g. is called the static margin. If the c.p. is behind the c.g., the static margin will be positive, and the aircraft will be stable. If the static margin is negative, the aircraft will be unstable. It is our purpose here to show that the value of the c.g. is a parameter which can be established as a requirements constraint.

The center of gravity is the point through which the weight vector for the whole aircraft passes. The c.g. is determined by the weight distribution of the individual components. Thus, if the c.g. for the whole aircraft is set at a predetermined location, then the location of the individual components will be limited. If the location of components cannot be changed to meet the c.g. requirements, then ballast must be added. Of course, ballast is always considered a last resort. A properly designed aircraft will need no ballast.

Moment of inertia

The moment of inertia is important in determining the forces required to roll or pitch the aircraft. For example, engines mounted nearer the fuselage will result in a smaller roll moment of inertia with a resulting requirement for less aileron torque to roll the aircraft.

5.3   Dimensions

Dimensions are any sort of dimensional limits which may limit the size of the aircraft or its subsystems. At the aircraft level, the wing span, for example, may be limited by airport facilities, such as the width of gates. At the subsystem level, the size of the equipment may be limited by the surrounding equipment or aircraft structure.

Like weight, dimensions may be limited by rail, road, or air transportation requirements. Dimensions also include dimensional tolerances, which is one requirement for the interchangeability of parts listed in Appendix 2.

5.4   Reliability

In the aircraft industry there are many requirements parameters included in the general category of reliability. Each of these may be specified in the requirements analysis. However, only two of these are considered important top-level requirements. These are dispatch reliability and operational reliability. The others are derived requirements: that is, they depend on some aspect of the solution for their value as discussed in Section 4.6.

Dispatch reliability is the pre-flight probability that the equipment will perform as specified within 15 minutes after being called upon to do so. Dispatch reliability is driven by the minimum equipment list (MEL), that is, the list of equipment which can be inoperable and still fly the aircraft safely.

The aircraft manufacturer establishes a master minimum equipment list (MMEL) from which the FAA reviews and approves. The airline selects a subset called the minimum equipment list (MEL). Hence, the airline may choose to delay a departure based on criteria more stringent than the FAA and tailored to the airline's airplane configuration and operation.

Operational reliability is the probability that the aircraft completes its mission. Both dispatch and operational reliabilities are parameters highly valued by the airline customers as discussed in Section 8.5.

The following reliability parameters are also used in the design of the aircraft. However, they all are derived requirements:

1. *LRU MTBF*   The line replaceable unit (LRU) mean time between failures (MTBF) is the average time in LRU hours between confirmed LRU failures.
2. *LRU MCBF*   The LRU mean cycles between failures (MCBF) is the number of operating cycles between confirmed LRU failures.
3. *LRU MCBUR*   The LRU mean cycles between unscheduled removals (MCBUR) is the average LRU operating cycles between unscheduled LRU removals.
4. *LRU MTBUR*   The LRU mean time between unscheduled removals (MTBUR) is the average time between LRU unscheduled removals.

Although most engineers understand the SE principle of allocation for parameters which are additive, such as weight, it is much more difficult to understand the principle of allocation when it is applied to reliability. Some reliabilities are additive; others are multiplicative. Appendix 1 provides a description of the allocation of multiplicative parameters, such as reliability, and shows some examples.

5.5   Human Factors

The EIA 632 (1999) definition of SE states that the purpose of SE is to evolve *"people, product, and process"* solutions. The establishment of requirements for people is one of the most difficult areas of SE. This difficulty is due to the vast complexity of human behavior and the inability of analysts to understand and quantify all of the independent variables that influence behavior. Chapanis (1996) examines human factors in an SE context and shows that, in spite of these difficulties, SE does indeed provide a useful set of tools for the establishment of human factors requirements.

Two views of humans in the system

People requirements can be viewed in two ways: as part of a system or as an interface with an aircraft component.

First, if people are viewed as part of the system, then SE can lay a limited number of requirements on the people. These requirements usually concern the number, training, and sometimes physical characteristics of the flight and maintenance crews. Chapanis (1996), however, cautions against relying too much on the model of the human as a system component because of the large number of variables which may affect human behavior. Nevertheless, Satchell (1993) cites an FAA study which showed that 33 percent of aircraft incidents (were due to "deviation from basic operating procedures." This is why the Commercial Aviation Safety Team (CAST) (2011) has placed such a high priority on improving operational procedures. Furthermore, many of these deviations resulted from training deficiencies. Hence, assigning requirements to the human, such as training requirements, is an essential part of the requirements process.

Secondly, people can be considered to be an interface with the aircraft components. It is this second view in which human factors can establish requirements for the components. Even this view of human factors is not without difficulties. FAR requirements will say, for example, that the fire extinguisher must be "readily available." It is the role of the human factors specialist to transform this statement into a verifiable requirement in accordance with SE principles. The final requirement may be, for example, that the fire extinguisher must be "no more than six feet from the oxygen mask." Of course, this requirement will have to be established by a human factors analysis and verified by examination as discussed in Section 11.2. Finally, it must be pointed out that the true systemist would hold that the view of the human as a system component is the correct one since the system is not *complete* without the human.

The following sections will discuss the five human interface types: flight crew, cabin crew, maintenance, ground service crew, and passengers. For each of these types, the human factors specialist will transform the human factors considerations into valid, verifiable SE criteria:

Anthropometry and strength

Anthropometry and strength pertain to how the system should allow humans to *fit* into it. Humans must be able to reach, stand, or sit where appropriate. Flight deck and other controls should allow movement within human strength capabilities. In most cases, dimensions should consider the fifth to the ninety-fifth percentile male and female population. All emergency equipment and exits, lights, and communications should meet the anthropometry and strength requirements which would enable the humans to use them effectively. For example, "the door handle shall allow operation with a force of 35 lb" is a clearly verifiable requirement.

Human factors analysis is critical in determining the reach and posture requirements for passengers, cabin crew, maintenance, and ground servicing. Three-dimensional computer models of both the human body and also the components to be maintained have been helpful in establishing these requirements. Finally, the study of human factors treats all aspects of passenger needs for controls, convenience, and comfort.

Vision

Flight deck vision requirements are concerned with the ability of the system to provide flight crew ability to see inside and outside the flight deck in accordance with FAR standards. In addition, the cabin crew should have a clear view of equipment and passengers both inside and outside the aircraft. Maintenance and service personnel should be able to see what they are working on. Passengers should be able to see relevant controls, indications, and safety equipment.

Ingress and egress

The human factors specialist has the responsibility for setting the requirements for flight crew ingress and egress during normal, abnormal, and emergency conditions. The flight crew should have egress through the window or flight deck door. Cabin crew should have egress through all normal and emergency exits. Maintenance and service personnel should be able to access all equipment and to reach service ports and stations. Passengers should be able to have ingress with carry-on luggage (under normal conditions) and have egress quickly (during abnormal and emergency conditions). These requirements should be specified in terms of the actual minimum physical dimensions and verified by examination.

Sound

The establishment of sound level requirements for human communication and stress management as discussed in Section 5.5 is an example of the necessity and value of the SE approach. This task requires that limits on sound levels be allocated across many subsystems to all potential sound sources, such as fans, air conditioning ducts, pumps, engines, and so on.

All sound information, such as alerts and communications, should be within human range and sufficiently above background noise levels to be understood. This information is used to establish the noise level limits discussed in Section 5.9. Maintenance and service crews should be able to communicate with others as necessary. Passengers should be able to communicate with each other and with the cabin crew. Overall, the aircraft environment should be free of annoying noise or pulse patterns.

Touch

Touch requirements pertain to the physical interaction of the human with the equipment. Touch temperatures should be below the value suitable for the material: that is, the temperature at which a material can be touched by a human is different for different materials. All rough or sharp edges should be eliminated. Controls (handles, knobs, and so on) should be similar for commonality and dissimilar enough to reduce confusion. Handle directions and emergency equipment should be intuitive. In some cases, in the flight deck, controls should conform in shape to the associated aircraft control surface. Equipment should be designed to allow maintenance and service personnel to work by feel. Passengers should be free from the dangers of sharp edges and be able to distinguish controls.

Cognitive considerations

Cognitive requirements—general

In general, cognitive requirements pertain to the way humans receive and act on information. These requirements are most important in the flight deck. Information should be provided to humans in such a way that they can understand and handle it with an error rate below some specified level. For the cabin and flight crews, memory requirements should be minimal, emergency operations should be intuitive, and workload and error must be minimized. Cognitive requirements also pertain to the training and psychological requirements applied to cabin crew members to enable them to maintain control of passengers in a variety of situations, such as during an emergency evacuation.

For all personnel (flight crew, cabin crew, maintenance personnel, service personnel, and passengers), equipment should be simple and intuitive.

Cognitive requirements—flight deck

The flight deck is the focal point of cognitive human factors requirements. Human factors analysis determines the requirements for flight deck information processing, the levels of automation, situational awareness, resource management, and procedures. Normal, abnormal, and emergency conditions should be analyzed in accordance with the situational functions described in Section 3.3. Particular attention should be paid to avoiding design induced errors and maintaining pilot vigilance. Vigilance is the ability of observers to maintain their attention and be alert to stimuli over prolonged periods of time. One area of particular focus is the design of cockpits which will enhance the pilot's situational awareness so that he or she will not become dependent on the aircraft's automated systems but will remain aware of hazardous situations. Human factors specialists endeavor to determine the root cause of accidents and establish requirements which will prevent their recurrence.

Although most aircraft accidents are attributed to human error, this fact does not imply that pilots are always culpable or even poorly trained. It implies that an aircraft needs to be designed to allow the pilots to handle the situations in which accidents are most likely to occur. For this reason, the role of the human factors specialist is important and the use of SE principles will aid in the task of establishing the requirements to meet this challenge.

Satchell (1993), for example, provides an analysis of flight deck requirements pertaining to resource management to avoid human error. Satchell suggests several functional requirements for the alerting system. For example, the alerting system "must measure the vigilance of the operator." This step is in clear agreement with the SE functional analysis process. However, the exact requirements for this measurement and the criteria for verification rest with the human factors specialist. He or she should use the human factors tool box of professional expertise, analysis, historical data, and simulation to establish these criteria.

From an SE point of view, the Navigate Aircraft function discussed in Section 3.3 can be allocated either to the flight crew or to the equipment. Section 9.2 discusses the allocation of this function to the avionics segment and the synthesis of the avionics equipment. From a human point of view, the subfunctions are more complex. In addition to the direct functions, such as Provide Control Commands, some additional subfunctions are discussed below.

A potential subfunction to be treated for the flight deck is Provide Peripheralization Modulation. Peripheralization is a complex psychological state which results from a shift in the pilot role from direct contact and control of the aircraft to one of system monitor. Satchell discusses the causes of peripheralization, primarily increased automation, and ways to modulate it. Peripheralization is a contributor to the loss of vigilance.

Another possible subfunction is Manage Stress. Stress is an emotional state, either detrimental or beneficial, which may affect flight crew performance. Stress may be caused by various factors, called stressors, such as temperature and vibration. Stress may be caused by outside factors, such as personality and cultural traits. Some stress factors may be static, others dynamic. Some stress may be beneficial; that is, it may provide a certain degree of alertness. Stress may result in arousal, that is, a state of alertness following termination of the stressor or perturbing event.

An obvious subfunction is Manage Workload. Shafer (1987) defines workload as the number of things an operator has to do within any particular time period modified by their level of difficulty. Workload assessment is a major human factors effort. A major requirement is to reduce the workload during periods of high activity, such as flying at low altitudes, and during emergencies. Human factors specialists have attempted to quantify workload difficulty with a measure called the Bedford scale. This scale helps human factors specialists create workload requirements which approach the SE goal of verifiability.

Human-centered Automation

Billings (1997) discusses the concept of human-centered automation. The premise of this concept is that humans must be the focal point of all requirements in the design of automated flight deck and air traffic control (ATC) systems. These requirements will augment the requirements for the avionics segment described in Section 9.2. The result of Billings' study is a set of requirements for both the flight deck and ATC systems. A typical requirement is that "designers should keep human operators involved in an operation by requiring of them meaningful and relevant tasks, regardless of the level of management being utilized by them." This type of requirement may not be easy to verify or to synthesize in accordance with SE principles. However, its articulation is a major step towards assuring that human requirements become the focal point of flight deck and ATC design. Chapter 6 on Interfaces discusses Billings' rules more extensively.

Equipment safety

Human factors analysis contributes greatly to the equipment safety analyses. Human factors safety analysis addresses the potential safety hazards which equipment poses to humans. Areas addressed include headstrikes, decompression, seat design for crash loads, corners, edges, and walking surfaces. For maintenance and service personnel, other considerations, such as fuel sparks and electrical shocks, are considered.

Conclusions

No other specialty offers such difficulties and opportunities for improvement in aircraft design as human factors. Difficulties arise from both the complexity of human behavior and the large number of variables involved. SE offers an integrated and methodical approach to addressing human factors within an aircraft context.

5.6   Environments

This category includes all environments which every component of the aircraft must endure; such as temperature, pressure, shock, vibration, and so on. The systems engineer should assure that all requirements are satisfied under the appropriate environments *and* combinations of environments. This category does not include any phenomena which the item might emit. For specifications, it is not necessary to specify the environment for each performance requirement. It is only necessary to have a single requirement which says that "the system shall meet its performance requirements during and after exposure to the following environments." Or, alternatively, the requirement might say that "the system shall meet its performance requirements *following* (as opposed to during) exposure to the following environments." Hence, it is important to specify the exact operational requirement relative to the environment.

It is up to the systems engineer to make sure that the requirements are verified in the appropriate environments. Although environmental analysis is an area which has always been important in the design of aircraft, SE brings to the table a methodology which recommends a much more rigorous incorporation of environments into the requirements process. That is, it recommends a multidimensional evaluation of environments: by operational phase, aircraft zone, the degree of abnormality of operation, and geographical zone.

In all cases it is important to specify quantitatively the range of environmental conditions, for example, temperature in degrees, sand in particles per cubic foot, and so on. Most of these requirements will be found in standard specifications.

Environments by phase

The critical feature of environments is that they constitute the conditions under which *all* performance requirements should be specified and tested. For example, as we have seen in Section 3.3, the environments should be established for each phase of aircraft operation, including maintenance and highway transport. For structural design, the FARs specify the conditions under which the critical environments occur.

Environments by zone

It is important to specify environments *by aircraft zone*: for example, in the passenger compartment, cargo compartment, wing, wheel-well compartment, and so on. Many environments are very zone dependent because they are attenuated by the aircraft itself. Hence, it will be impossible accurately to estimate the environments in those zones until after an initial concept is formulated. An example of such a requirement identified after concept development is noise which is greatly attenuated by the body of the aircraft. Zonal environments are derived since they depend on aircraft design for their values. Hence, zonal environments could be identified no sooner than the system design review (SDR) discussed in Section 12.4 and no later than preliminary design review (PDR) also in Section 12.4.

Environments by normal, abnormal, and emergency operation

Included in these functions are the normal, abnormal, and emergency aircraft operations. The functions force the requirements analysis to examine all environmental conditions any element of the aircraft may encounter. Cabin depressurization is an example.

Environments by geographical region

Environments for specific customers can often be more severe than for most other customers. In that case, it should be assured that the aircraft will withstand the environments of long heat soaks in the desert, for example, or for long cold exposure in the Arctic. Similarly, the desert environment will have other severe aspects, such as increased levels of sand and dust. Geographical environments are requirements at the top level of the aircraft hierarchy since they do not depend on the design. Hence, geographical environmental definition would be presented at the SDR.

Environment types

Following is a list of environments which should be specified for an aircraft:

Temperature
For exterior components, the temperature environment will be determined by the altitude temperature profile and aerodynamic heating. In addition, on the aircraft there is considerable *induced* heating, for example, for components in the vicinity of the engine. Temperatures on the ground will be determined by the *heat soak*, that is, for aircraft which have been sitting on the runway for hours on a very hot day.

Pressure
For exterior components, the pressure will normally be determined by the atmospheric pressure and the aerodynamic pressure profile on the aircraft. For internal components, the pressure will be either in the pressure shell, or outside. Among the emergency operations is the decompression condition. During decompression, not only should pressure be specified, but also the pressure *rate*. The airframe's ability to maintain the cabin pressure is considered a performance requirement rather than a constraint.

Electromagnetic interference (EMI) and high-intensity radio fields (HIRF)
EMI is always an *induced* environment, that is, it is an environment produced by some other aircraft component, namely, electrical components. In addition, EMI can be produced by on-board equipment carried by passengers, such as portable computers. Electronic equipment is the primary type of component susceptible to EMI. Hence, the EMI environment is the environment for which each piece of electronic equipment should be designed. In addition, the equipment should be located no nearer than specified distances from electrical cables.

   HIRF is similar to EMI except that it is an externally generated, normally ground-based, electromagnetic environment the aircraft may be subjected to.

Shock, vibration, and load factors
Shock is the result of major impacts, such as hard landings or ground impact. Shock environments will be specified *as they are transmitted* by the aircraft structure. Similarly vibration can result from engine or auxiliary power unit (APU) operation or the result of aerodynamic vortices. Like shock, vibration will be transmitted by the aircraft structure. Thus, for both shock and vibration, the

environment for each aircraft component will depend on its location in the aircraft and the phase of flight.

Shock and vibration can also occur during transportation to the final assembly location, during final assembly, and on rough ground taxiways and runways. Hence, these phases should also be considered in the determination of the shock and vibration environments.

Load factors are those loads which are not of a fast, impulsive nature. These are the g-loads experienced by the aircraft in flight and upon landing. Although shock, vibration, and other load factors may be applied to any component, they are the primary environments against which all structures are designed.

Lightning

All equipment is normally subject to the requirement to withstand the effects of single- and multiple-stroke lightning. These effects are most critical for fuel lines, electrical conduits, and other lines that pass through wing tanks. Line connectors that are not totally conductive are subject to sparking and may result in the inflammation of the vapors in the fuel tank. Newer aircraft suppress this effect by inserting nitrogen enriched air (NEA) into the vapor region.

The lightning requirement for individual connectors is normally of the form <The connector shall be spark-free when subjected to a current of TBD amps.> Of course, TBD (to be determined) will vary from connector to connector throughout the aircraft. The value of TBD can be determined by *deriving* it from the basic input charge for the whole aircraft, which may be, for example, 200,000 amps. This is an example of a derived requirement as discussed in Chapter 4.

This requirement can be verified by passing a current through a connector in a laboratory and observing a resulting spark, if any. If there is no spark, the requirement will have been met. This is an example of verification by test as described in Chapter 11.

Sand and dust

Normally, sand and dust are only encountered when the aircraft is on the ground. During run-up, taxiing, or take-off, sand and dust are only likely to affect engine operation, the exterior of the aircraft, and the intake air. In a ground maintenance, storage, or loading condition, it is important to specify the sand and dust as they might affect cargo loading equipment. Hence, every phase of the flight profile is important. Additionally, if the support equipment is part of the aircraft *system* as discussed in Section 2.3, then the effect of the sand and dust on that equipment should be considered also. Furthermore, the effect on interior systems should be considered when the aircraft is on the ground with the doors open.

Sand and dust may enter the aircraft air conditioning system through the bleed air system especially when the aircraft is operating in a particularly severe environment, such as in a desert climate. In these instances extra filters are often added to the bleed air system.

Fungus
Aircraft materials should be free from damage by various types of fungus since many carriers operate in regions of the world where fungi are abundant. Lists of fungi are available through standard sources.

Corrosive environment
This pertains to any sort of toxic environment which the aircraft materials may come in contact with, including human perspiration, while the aircraft is being flown or maintained. More severe toxic environments include acids or hydraulic fluids. This category also includes salt spray. Corrosion protection is an FAA item for continued attention.

Solar radiation
Solar radiation is primarily an environment which is important when the aircraft is on the ground. Solar radiation includes both the spectral distribution and level of solar radiation the aircraft will be subjected to. It not only affects the exterior of the aircraft but also is the input heat load for calculating the maximum interior temperatures on the ground.

Humidity
Most aircraft components are designed to a humidity range from 0 to 100 percent.

Precipitation
The precipitation requirements specify the environments for rain, snow, hail, frost, and sleet which the aircraft must withstand. Snow requirements include snow loads on the aircraft.

Foreign object debris (FOD)
FOD is the environment of objects which may be on the runways of various airports and may damage the aircraft, especially the engines.

Noise
This is the noise each aircraft component must *endure* as opposed to the noise it emits as explained in Section 5.9. This environment would be particularly important to components within the engine nacelle, for example.

Volcanic ash
The attention to volcanic ash has increased since the eruption of the Icelandic volcano Eyjafjallajökull in 2010. At a meeting of the International Volcanic Ash Task Force (IVATF) in 2010 the FAA presented a list of ways to mitigate the effects of volcanic ash. According to EasyJet (2014) following the eruption of 2010 development has begun on a device called AVOID (airborne volcanic object identifier and detector) that would detect volcanic ash and allow the pilot to avoid

the volcanic ash area. Implementation of this device would be an example of employing the *drift correction* resilience principle described in Chapter 16.

5.7 Maintainability

The concept of maintainability includes both the constraints on the item to be maintained (the aircraft) and the requirements for the maintenance equipment (ground support equipment). Hence, maintainability spans the elements of the aircraft system.

SE brings two concepts to maintainability which promise to improve the quality of the aircraft system. First is the concept of verifiable requirements. Maintainability has traditionally been thought of as a discipline not amenable to verifiable requirements. We have seen that the role of the systems engineer is to convert the qualitative requirements into verifiable requirements. Secondly, SE helps build maintainability into the aircraft from the inception of the program.

Quantitative maintainability requirements

Many maintainability requirements are quantitative and are therefore verifiable.

1. *Maintainability cost* The principal requirements parameter for maintainability is maintenance cost per 1,000 flight hours (MN$/1000FH). This parameter can be converted into maintenance man-hours per 1,000 flight hours (MMH/1000FH). Material cost per 1,000 flight hours (MT$/1000FH) can also be separately specified.
2. *Fault isolation* Specific and verifiable time limits can be established for both on-aircraft and off-aircraft fault isolation times. On-aircraft repair time is measured by the mean time to repair (MTTR). A basic maintainability requirement for both safety and economic failures is that "all failures shall be evident" or not affect safety, economics, or operations (for any subsystem or for the aircraft as a whole). *Evident* can either mean that there is an indication to the crew during the course of their normal duties (alert, light, and so on) or that it would be evident to an observer, such as to a person doing an aircraft walk-around during turnover.

Qualitative maintainability requirements

Qualitative maintainability covers many areas: accessibility, interchangeability, off-the-shelf components, overhaul, component wear, standard tools, handling equipment, support equipment, erroneous installation, contamination, identification tags and labels, repair procedures, corrosion. It is necessary to express all qualitative requirements in verifiable terms.

Safety-related maintainability

Safety-related maintainability is of primary importance. Many of the items discussed above, such as fault isolation requirements, are safety related. These items will be part of the certification data package.

Maintenance types

Maintenance costs are incurred during six types of maintenance: overnight checks, A checks, B checks (rarely used), C checks, unscheduled maintenance, and fixed interval checks. The overnight check is done when the aircraft is in service at any airport and consists of routine tasks, such as fluid checks. The A checks occur at predetermined intervals and consist of preventive maintenance tasks. B checks and C checks are at incrementally longer intervals than A checks. Each level of maintenance type is progressively deeper in its diagnostic inspections. Although it is a basic goal of maintainability to minimize the scheduled maintenance, it is never a practice to *require* scheduled maintenance. Unscheduled maintenance is driven by the MTBUR.

Fixed interval maintenance

Fixed interval maintenance is also undesirable and is maintenance mandated at intervals different from the regular scheduled maintenance. It is a maintenance cost driver and results from FAA mandated certification maintenance requirements (CMRs). A CMR is a required periodic task established during the design certification of the aircraft as an operating limit. It is intended to detect hidden faults, that is, safety-significant latent failures which would, in combination with one or more other specific failures or events, result in hazardous or catastrophic failure conditions.

The way to reduce the number of CMRs is to do a thorough fault tree analysis and to pass the system safety analysis (SSA). These steps may result in requirements for increased redundancy and reliability and a reduction in non-evident faults.

Accessibility

We saw above that accessibility was listed as a qualitative requirement needed to be turned into a verifiable requirement. There are two ways to do this: The first is to establish a time limit on the removal and replacement of an LRU. These limits can easily be verified by demonstration. The second is to establish, during the design phase, dimensional criteria for accessibility: for example, the space between the structure and an LRU should be at least TBD inches for removal and replacement. This requirement can be verified by examination.

Trade-offs between accessibility and reliability are a necessity. For example, if the MTBF of a component exceeds the planned life of the aircraft, then its accessibility is not so important. On the other hand, components with low or unknown reliabilities need to be extremely accessible.

Spares

The cost of spares at the time of purchase of the aircraft is a major factor in the purchase of aircraft. Reduction in the number and cost of spares would significantly aid in the sale of aircraft. The following items drive the number and cost of spares. If these items are addressed during the requirements phase, then SE can be a major help in the reduction of the number and cost of spares:

1. *MEL* A goal of maintainability is to put all aircraft components on the MEL. A longer MEL will reduce the number of spares. The best way of getting an item on the MEL is to build redundancy into the system. This can be done by creating an up-front requirement for redundancy. Of course, all items cannot be redundant. Trade-offs of redundancy against cost and reliability are necessary. It is also possible to make a subsystem fault tolerant by assuring that minor failures do not render a larger system inoperative.

2. *MTBF, MTBUR, and MTTR* It is also possible to specify the required values for these parameters, which can be built into the item. Another maintainability goal is to make MTBUR equal to MTBF, that is, to remove all LRUs only at the scheduled rate. There are two ways to do this: design foolproof fault detection procedures, and make built-in test equipment (BITE) 100 percent reliable.

3. *Non-interchangeability* It is possible to specify the interchangeability of an item, either with other aircraft models or with other items on the same aircraft. This type of requirement is a standard part of specifications (Appendix 2). Interchangeability does not necessarily mean that interchangeable items are identical or made by the same supplier. Interchangeability can be achieved by assuring that their interfaces, both functional and physical, are identical (Chapter 6).

4. *Sole-sourcing* This is an economic factor which is part of the supplier management process. It means the awarding of a contract to a supplier without a competition. Competition among potential suppliers will assure both lower aircraft costs and lower spares costs. The manufacturer should consider the possible consequences of sole-sourcing as part of the risk management process discussed in Chapter 15.

5. *Number of line stations and transit time* The aircraft manufacturer has little control over the number of airline hubs and the distance between them. Nevertheless, they can be considered to be part of the aircraft system in a larger sense. The airline should consider the effect of these factors on spares costs when establishing their line maintenance within their route structure.

Building maintainability into the aircraft

Maintainability is an area in which integrated product development (IPD) multifunctional teams are most valuable. This value results from the fact that an organization's maintainability department may be organizationally and physically separated from design engineering. The IPD multifunctional teams encourage the maintainability engineers to be an integral part of the requirements and design process described in Section 12.3. In this way maintainability requirements can be incorporated into the design before concepts are formulated. That is, in the SE process the design is not evaluated for maintainability after it has been formulated, but rather the maintainability requirements are incorporated with all other requirements.

5.8   Design Standards

This is another category which would include things that appear in company, industry, or regulatory standards, like design margins, prohibited materials, and so on. Companies normally have well-documented design standards in each design specialty.

5.9   Emitted Noise

This is the limitation on noise that the item is allowed to emit. It is not the noise environment described in Section 5.6. Many engineers get these two items confused. It is important to distinguish between what the item must do (limit its noise output) and what environment it must withstand (noise). We saw in Section 4.7 that the emitted noise limit can be allocated to the various aircraft elements. Of course, noise is highly frequency dependent. The noise limitations in each frequency band should be specified. The two primary noise emitters are the aircraft engines and the fuselage boundary layer. However, many other components are capable of emitting noise, such as air conditioning ducts and hydraulic pumps. Hence, noise emission limits should be laid on all components.

For noise from external sources, there are three possible ways of limiting the noise in the cabin: limiting the source (engine or boundary layer), insulating the cabin, or actively suppressing noise. Trade-offs will determine which of these is most effective.

5.10   Emitted Electromagnetic Interference (EMI)

EMI is a major consideration in aircraft design. All electrical components should be shown not to emit more than a fixed amount of EMI which may interfere with avionics operation.

This is the EMI that an item is allowed to emit and not the EMI that an item must endure. However, it is necessary to consider the emitted EMI constraint jointly with the EMI environment on the vulnerable component, normally an electronics item. Emitted EMI can be constrained, to a certain extent, by insulation. However, the electronics component should be, at the same time, protected by separation from the emitting cables and by shielding.

5.11   Cost

Cost is an important constraint, especially in the aircraft industry. We will see in Section 8.6 how the entire aircraft is designed with cost constraints in mind. In DFMA projects, for example, items are redesigned with a specific cost reduction goal in mind. However, if we design the aircraft *in the beginning* with DFMA in mind, that is, with specific cost constraints, then DFMA will never be needed. To do this, it is necessary to consider as a major cost driver the assembly costs, which are in turn, to a major extent, driven by the number of parts. Hence, parts reduction will be a major consideration in minimizing cost. We saw before how cost can be allocated to the various aircraft elements. Normally, in design-to-cost, the principal cost parameter used is recurring (unit) cost. However, the total direct operating cost (DOC) and non-recurring (development) cost can also be used.

5.12   Transportability

Transportability includes all the design limitations on an item from having to be transported. These limitations can be included in weight, dimensions, shock, and vibration. For example, if the fuselage is being assembled by a supplier and transported to the manufacturer's facility by truck, the size of the truck will limit the size of the fuselage segment. Therefore, we should design the fuselage to be transported in segments small enough to be transported by truck. Although transportability is not an issue of major concern in aircraft design. Nevertheless, transportability requirements should be considered whenever they apply.

5.13   Flexibility and Expansion

It is often a design requirement on systems, such as aircraft, to design them for growth. That is, it may be necessary to design them with features, such as a larger wing, which may be needed on *future* versions or derivative models described in Section 2.2. For example, if a future aircraft model is expected to carry a larger payload, then the current aircraft should be designed to meet the future aircraft's performance capability. The wings and propulsion system should be selected, for example, to meet the future range, speed, and cruise altitude requirements.

5.14   Producibility

EIA 632 (1999) states that SE evolves life-cycle solutions for "people, product, and *processes*." It is the process aspect that the producibility requirements address. If two airframe components, for example, the landing gear door and the wing, are expected to meet with a certain tolerance, then requirements should be laid on both the component design and the manufacturing process. First of all, the component mismatch requirements are derived partly from the Generate Aero Forces function. These requirements will establish the limits on the aerodynamic drag caused by the mismatch. A manufacturing analysis will establish the derived requirements for the tolerances in mismatch resulting from the tool design and assembly sequence allocated to both the door and wing assembly processes.

6

Interfaces

The greatest leverage in system architecting is at the interfaces.

Eberhardt Rechtin

One of the goals of SE is *completeness*, that is, to make sure that every aspect of the system has been covered and incorporated into the design. An interface is a boundary between two system elements. Yet it is much more; interfaces are one of those completeness areas which, in traditional engineering, suffer from insufficient treatment, with resultant risk and possible harmful consequences. Alexander (1964) says that a good system is one with the fewest number of misfits. In addition, Rechtin (1991, p. 29) points to the importance of interfaces in the architecting of a system. But, of course, we know that any system should meet its top-level requirements as well, as discussed in Section 4.2.

The purpose of this chapter is to give some basic guidance regarding interfaces, especially pertaining to commercial aircraft, to capture the key aspects of interfaces, and to incorporate them into the design.

There are two types of interfaces: functional and physical. We should, therefore, understand each type of interface and show how interfaces fit into our design.

The importance of interfaces

Interfaces are important from two perspectives, first, a *holistic* perspective of a system, and secondly the system property of *cohesion*. Chapter 4 explains that the requirements for the elements of a system should be determined holistically rather than from a reductionist point of view.

The Vee model explained in Chapter 4 is a top-down, vertical determination of requirements. As pointed out in Chapter 4, this method, used alone, fails to determine the relationships between the elements. Interface requirements are the requirements *between* the elements. If there is a conflict between these two methods, a trade study must be conducted. Chapter 4 also explains requirements trade studies.

In addition, as explained by Hitchins (1993, p. 55) a basic property of a system is *cohesion*. This simply means that for a system to be a system there must be a relationship between the parts. Interfaces define that relationship between the parts.

6.1   Functional Interfaces

Functional interfaces are the most neglected type of interface. And, yet, they are the most important because they characterize the whole purpose, that is, the function of the interface. With the goal of *completeness* in mind, it is well to remember that there are at least two associated functions for every interface.

Since all performance requirements are traceable to functions as explained in Section 4.2, it follows that there will be at least two requirements associated with each interface. Figure 6.1 illustrates this idea:

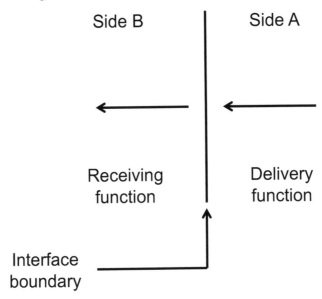

Figure 6.1   Interface functions

We see in this illustration that Side A delivers some quantity, say electrical power, to Side B. At the same time Side B receives that same quantity, electricity, from Side A. The two implied requirements are:

1. Side A must generate the quantity, and
2. Side B must perform a function with the quantity that Side A delivered.

As we saw in Section 4.1, these requirements should be expressed in verifiable terms. For example, "the [Side A] subsystem shall deliver electrical power at 28 volts," as expressed in strict requirements terminology.

The quantity on Side B is known as an "antecedent," (Carson, 1995). That is, Side B performs a function *using the quantity provided by Side A*. For example, say the element of Side A is a fan. Then, the requirement for Side B might be: "The fan [Side B] shall provide 20 cfm of air, using the 29-volt power supply of the electrical system [Side A]."

Hence, we have developed a completeness paradigm. It is a check on the requirements analysis of Chapter 4. If the requirements analysis was performed on this hypothetical fan and failed to identify the two requirements we just identified, then the analysis was not complete or thorough. Now that we have reviewed what functional interfaces are, let's examine a few common ones in commercial aircraft. Remember, each of the parameters listed should be quantified and converted into a *bona fide* requirement as discussed in Section 4.1.

Electrical power interfaces

Electrical power is one of the most common interfaces on the aircraft. However, most electrical interfaces are internal to the electrical subsystem and, therefore, are not subject to the strict interface management that interfaces between elements are. However, the electrical subsystem does not "own" all the electrical elements. For example, the hypothetical fan, discussed above, may belong to the environmental control subsystem (ECS). The typical characteristics (functional interface parameters) of the electrical power are: voltage, current, alternating current (AC) or direct current (DC), power, phase, and high and low values of the power, voltage, and current.

Hydraulic power

Like electrical power, the hydraulic subsystem only has an interface with another element when the other element owns a component utilizing hydraulic power. The primary hydraulic characteristic is hydraulic pressure.

Pneumatic power

The pneumatic subsystem may provide air to the ECS and provide cabin pressure. Pneumatic air can also be used to start the engines and to de-ice certain exterior components. Typical pneumatic characteristics include pneumatic pressure and temperature.

Mechanical forces and torques

Mechanical forces and torques constitute the interface functions between many aircraft elements. For example, the control system provides mechanical forces to move the control surfaces, such as ailerons and elevators. The engines provide mechanical torque to run the electrical generators. These are only two among many.

Conditioned air

Conditioned air passes across many boundaries: from the ECS ducts into the cabin, from the cabin into the lavatories, from the cabin into cargo area, from the air ducts into avionics racks, for example. The characteristics of the air are well-known: flow rate, temperature, humidity, quality (absence of particulates and ozone), and pressure.

Heat

Heat is a special case because it comes in two forms: desirable and undesirable. Desirable heat travels across the interfaces between heat exchangers and the fluids, either receiving or delivering the heat, for example. One example of desirable heat exchange is the cooling of engine bleed air.

It is just as important to capture and characterize the undesirable heat transfer across interfaces. This heat transfer is equivalent to a local adverse temperature environment. It is also advisable to cross-reference any adverse environments as part of the local environments in the environmental analysis described in Section 5.6. As an example, say that a component, a duct, for example, is attached to the engine by a brace. Depending on the material it is made of, this brace will transmit heat from the engine to the duct. Considering the high-heat environment of the engine, it is necessary to minimize this heat transfer from the engine to the duct so as not to cause damage to the duct.

Vibration, shock, and loads

Like heat, vibration, shock, and loads, in general, constitute undesirable interfaces. Aerodynamic loads on the structure are not interfaces since the air is not a system element. Take the hypothetical duct connected to the engine in the example above. The brace holding the duct will transmit vibration to the duct in the same way it transmitted heat. Hence, the vibration transmitted is a functional characteristic of the interface which needs to be recorded and incorporated into the design. Furthermore, vibration will be transmitted throughout the entire aircraft through the structure and transmitted to all components with which the structure interfaces. At each interface any transmitted vibration, shock, and loads need to be captured and analyzed. The characteristics of these parameters are well-known. The primary ones are magnitude and frequency.

Signal interfaces

Signal interfaces, also called information interfaces, carry the vast amount of information that is transmitted over the aircraft and between the aircraft and the ground, satellites, and other aircraft. Like electrical power, signal interface parameters include voltage and current, for example. In addition, various signal protocol properties should be specified. Example signals include warnings and fault detection signals from the various subsystems, such as the propulsion subsystem.

To facilitate the integration of aircraft avionics, the Society of Automotive Engineers (SAE) has developed a standard, AS4893 (1996), which defines a set of generic interfaces. The purpose of this standard is to increase the chance that components produced independently will have compatible interfaces. In addition, it provides a basis for commonality for both vendors and users of these components.

6.2 Physical Interfaces

The purpose of the physical interface analysis is to assure that the physical configurations of both system elements are compatible; that is, they fit together. Physical interfaces are one area that is traditionally well handled. However, mistakes do occur and improvement is needed. Two things are needed: First, rigorous SE documentation will assure that all physical aspects are considered. Secondly, new technological advances, such as the electronic developments fixture (EDF) will make it easier to capture the complete physical aspects of the interface. The EDF is discussed below. The interface control drawing (ICD) is the primary tool for recording the physical interfaces.

6.3 External Interfaces

Many aspects of external interfaces are often the most neglected areas of aircraft design. In the context of the entire aircraft external interfaces include communications, service, maintenance, and facilities. In the context of a subsystem or other aircraft element, *external* means external to that element. For example, the external interfaces to a lavatory include the electrical connections needed for lights and fire detection, the plumbing interfaces for water and waste, the air interfaces for environmental control, and the physical interfaces with the surrounding aircraft structure. However, we will focus on interfaces external to the aircraft here.

Service

Aircraft service interfaces are one area for potential neglect since they are often not considered at the beginning until the aircraft design is well under way. Service interfaces include food service, cleaning, cargo loading, fuel loading. In general the service equipment exists and is not subject to redesign for a specific aircraft. However, service equipment may vary widely among airlines and airports. It is advisable to have a complete physical and functional description of all service equipment used by the airline customers and airports. Hence, it is necessary to take all of these variations into consideration. In addition, as part of the design of new types of aircraft, such as the high-speed civil transport (HSCT), the radical changes in configuration may impose new service requirements with resulting impacts on both the aircraft and the service equipment.

Maintenance

The conclusions for maintenance interfaces are generally the same as those for service above. In short, it is necessary to consider the interface of maintenance equipment with the aircraft the same as any other interface. Like service, airlines vary greatly in the maintenance equipment they use. It is desirable to obtain the characteristics of this maintenance equipment because it will affect the design. This list will include all the types of equipment used to test the aircraft, all the heavy-lift equipment, such as fork-lifts, and small tools. All of this equipment constitutes equipment with which the aircraft will interface.

Facilities

Facilities are also a neglected interface area. The passenger access tunnel, commonly called the jetway, interfaces with the aircraft on the ground. It is imperative to consider this fact in both the design of the aircraft and the jetway so that passengers are loaded and loaded optimally and so that no damage is done to the aircraft during that operation by the jetway.

6.4   Internal Interfaces

We have already discussed the types of internal interfaces (to the aircraft) in Section 6.1. The main idea regarding internal interfaces is that all external interfaces at any level of the aircraft hierarchy are internal interfaces at the next higher level. This is a very important and practical idea. However, in order to make sense, it should be combined with the concept that each level of the aircraft hierarchy has an owner. For example, consider the diagram of Figure 6.2.

This figure shows three levels of the aircraft system hierarchy. At the top level the aircraft interfaces with the maintenance system at interface "A." This interface is *external* to the aircraft itself. At the second level the ECS interfaces with the

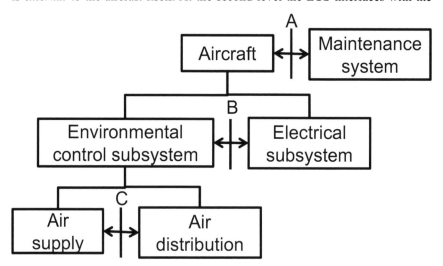

Figure 6.2   Interfaces in the aircraft hierarchy

electrical subsystem at interface "B." Interface "B" is, then, *internal* to the aircraft but *external* to both the ECS and the electrical system. This tiering of external and internal interfaces continues to the third level and Interface "C" which is internal to the ECS but external to the air supply and the air distribution subsystems.

The importance of this external/internal interface distinction has to do with the level ownership as discussed above, and with the verification of the requirements associated with each interface. Say, for example, that the aircraft program manager owns the top level, the ECS manager owns the ECS, and a supplier owns the air supply. We also learned in Chapter 1 that system verification is bottom-up. All this means that, first, the air supply would be integrated with the air distribution before it is integrated as part of the ECS. The ECS owner would have responsibility for assuring that the level 3 verification is successful. Similarly, the aircraft owner would assure that the level 2 verification is successful. Hence, verification and integration rise to the top with the ownership hierarchy like bubbles in champagne.

6.5 Operational Interfaces

Operational interfaces have to do with the interfaces between any two elements of the system, either the aircraft system or external systems, during the operation of the aircraft. These interfaces can be either external or internal.

External interfaces

One of the primary external operational interfaces is communications. Communications are one of the few areas in which no physical interface exists. However, this fact does not diminish the importance of the functional aspects of external communications. The requirements for external communications with airport towers, other aircraft, and other ground nodes are generally well-known and documented.

Other operational interfaces according to the *FAA Systems Engineering Manual* (2014, pp. 57, 67) are the external interfaces, such as navigation and air traffic control (ATC). Factors to be determined through these interfaces are traffic density, flight phases, route configuration, visual flight rules (VFR), and instrument flight rules (IFR).

Human-automation interfaces

One of the most comprehensive sets of rules for human-automation interfaces is the one compiled by Billings (1997, pp. 237–246). Following is a summary of those rules:

- the human operator must be in command;
- to command effectively the human operator must be involved;

- to remain involved the human operator must be appropriately informed;
- the human operator must be informed about automated systems behavior;
- automated systems must be predictable;
- automated systems must monitor the human operator;
- each agent in an intelligent human–machine system must have knowledge of the intent of the other agents;
- functions should be automated only if there is good reason for doing so;
- automation should be simple to train, learn, and operate.

6.6　Interface Management

The very same owners at each level discussed above have another role. That is, one group, department, or company is responsible for assuring that all the features of its side of the interface are correct; that is, each party has designed into its side all the features to make the interface work properly. Typical owners are the manufacturer and the supplier. It is the responsibility of the interface management function to assure that this ownership principle is in place, that both owners recognize their responsibilities, and that they have incorporated the interface characteristics into their design.

It will be a key function of SE management as described in Section 12.1 to schedule and organize the interface meetings between the two owners of the sides. The product of the interface meeting is a simple agreement, sometimes called a *scope sheet*, which does the following:

1. describes the hardware owned by each side;
2. describes the functions received or delivered by each side;
3. obtains the signed concurrence of both sides to the agreement.

The scope sheet can be a simple document of no more than a single page to achieve its purpose.

6.7　The Interface Control Drawing (ICD)

The ICD is the basic product of the interface task. In its final form the ICD is a clear documentation of the interface definition. It shows clearly which items belong to each side. It shows the exact values of the interface functions and their tolerances. However, what may be surprising to some is that the ICD is not used to design any part of the aircraft. That information is on the drawings of each component involved in the interface. The requirements associated with the interface functions are located in the performance sections of the specifications. Hence, the ICD serves as a good coordination document.

6.8   Development Fixtures (DFs)

A development fixture (DF) is a mock-up of the aircraft used during development to assure that the spatial allocation for all components is correct and that they fit correctly. Traditionally, DFs were hard; that is, they were actual physical models of the aircraft, made of wood and other materials. As engineers developed each component or subsystem, they would place a physical mock-up of the part into the DF.

Recent computer technology has made possible the use of the electronic development fixture (EDF). That is, engineers create a three-dimensional model of the entire aircraft on a computer. The EDF makes possible the allocation of space for a component by reserving a zone for it. In spite of large company investments in computers and training required for EDF capability, the benefits are enormous. First, the EDF results in a significant reduction in lead time for development. Secondly, the cost savings resulting from the elimination of the hard DF will go a long way towards offsetting the EDF costs.

EDF has a significant role in the SE interface process. First and foremost, the EDF model will become an integral part of the ICD. Each party to the interface will use the EDF, rather than the traditional two-dimensional drawing, to control the physical interface. Hence, the EDF will add significant value to the interface process.

6.9   The N² Diagram

The N^2 diagram illustrated in Figure 6.3 is a useful tool in SE. It assures that all the functions identified in the functional analyses are reflected in functional interfaces. Each node in the N^2 diagram indicates a possible functional interface. Notice that in the example the Provide Electrical Power and Provide Environmental Control provide power and cooling to the other functions. However, the Navigate Aircraft function does not provide any quantity to the other functions.

In addition to being a useful tool, the N^2 diagram is one of the easiest to implement. Almost any word processing or spreadsheet program can be used to generate it. No special applications are required.

6.10   Interface Requirements

Interface requirements are the requirements associated with the interface functions discussed in Section 6.1 and the physical interfaces discussed in Section 6.2. The certification process requires as discussed in Section 4.9, Item 8 that interface requirements be provided as part of the requirements documentation portion of the certification data package.

	Provide electrical power	Provide environmental control	Provide guidance & navigation
Provide electrical power		Provide power to ECS	Provide power to avionics
Provide environmental control	Cool electrical components		Cool avionics
Provide guidance & navigation			

Figure 6.3   N² diagram

6.11   Interface Verification

As shown above in the discussion of internal interfaces, it is not possible or meaningful to verify interfaces, as such. Rather, the correct step is to verify the requirements associated with the interfaces. In addition, during system build-up, it is necessary to verify the functionality of the lower-level elements as a part of the verification of the higher-level elements, as illustrated in Figure 6.2.

7

Synthesis

Aircraft synthesis is the actual act of designing the aircraft or a segment of it. In the previous chapters we have said nothing about designing the aircraft. That is the basic point of SE: namely, that the functions, performance requirements, and all the constraints will have been so well defined that the design will now be much easier.

We have also seen that the first steps in aircraft system synthesis began when the aircraft architecture described in Section 2.3 and the aircraft system functions in Section 3.2 were defined. Synthesis also occurs when a requirement is allocated (definition 2—Glossary) to an element of the aircraft architecture. The final step occurs when the architecture, functions, and requirements are converted into a design. Hence, synthesis is a collection of steps which occur throughout the SE process. Figure 7.1 illustrates the synthesis process through the path between the hierarchies of functions, requirements, and component architecture.

It must be remembered though that the hierarchical path of Figure 7.1 is only part of the process. The vertical process shown in this chart is the *reductionist* path as explained in Section 4.3. The inclusion of requirements from other sources, such as operations, maintenance, and so forth are required to make the synthesis process *holistic*.

System synthesis has traditionally been considered the domain of the design engineer as distinct from the systems engineer. It is a common misperception that systems engineers overly constrain the design and stifle the creativity of the design. On the contrary, it is the job of the systems engineer to define the system goals and the conditions under which the system must operate so that the designer is free to create the best system possible. Ultimately, though, the ideal state is that SE principles will be so well understood by all engineers that this distinction will no longer be relevant and that SE can be used to enhance creativity rather than stifle it.

It is not the goal of this chapter to present the detailed steps and data to design an aircraft. Such sources as Corning (1977) do an excellent job of that. It is our goal to present the basic aircraft synthesis parameters and to answer the questions: Which parameters are performance requirements? Which parameters are constraints? And which parameters are design solutions subject to trade-off? Chapters 8 and 9 will answer these questions. This chapter discusses synthesis in the overall SE process.

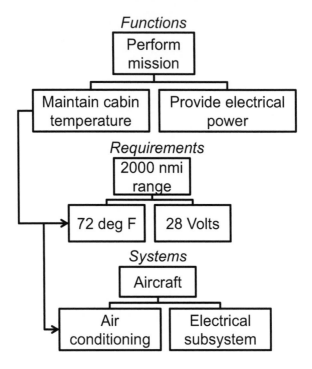

Figure 7.1   The synthesis path

Introducing Holism in the synthesis process

We saw in Section 4.3 that a key principle of SE is *holism*, that is, the creation of a system as a whole and not by looking at the individual parts. The latter is called *reductionism*. If the designer looked at only the requirements flowed down to the individual parts using the Vee model described in Section 4.3, the result would be a reductionist process rather than a holistic process. To obtain a holistic view of the system, the designer needs to look at the requirements from different sources that may be in conflict with the flowed down requirements. These requirements may be from production, operations, maintenance, or any of the other sources discussed in Chapter 4, for example human factors. Section 4.8 also provides several ways the designer can conduct trade-offs among these conflicting requirements. In the end the designer will have requirements for the individual components now holistically determined and not reductionist.

7.1   Aircraft Architecture

We have seen in Section 2.3 that one of the first steps in synthesis is the creation of the aircraft architectural hierarchy shown in Figure 2.1. For a new aircraft in Section

2.2, it is necessary to create and refine the hierarchy throughout the development of the aircraft. For a derivative aircraft also in Section 2.2 or change-based aircraft also in Section 2.2, the architecture will already have been defined. However, elements may be added to or deleted from the original architecture as the new design demands.

7.2   Initial Concept

The first step involving the conceptualization of real hardware and software is the initial concept. For new or derivative aircraft in Section 2.2, this concept would involve the development of a complete aircraft concept, its dimensions, weight, and performance estimate discussed in Section 8.2. This concept is based on the top-level (not derived) requirements. For derivative aircraft, the initial concept would specify major components, for example, an entire wing, fuselage, or tail from a previous design which would be employed. For change-based designs described in Section 2.2, the initial concept would also specify major dimensions, weights, and off-the-shelf items.

The key aspect of the initial concept is that it is *initial*: that is, it is subject to changes based on the trade-offs to follow. The key milestone for the initial concept is the system design review (SDR) described in Section 12.4. The purpose of the SDR is to show that the initial concept meets the top-level requirements and that the trade-offs have been defined which will result in a final baseline configuration at the preliminary design review (PDR) described in Section 12.4.

7.3   Trade-Off Studies

The trade-off process is a key element of SE when applied to aircraft. A trade-off is an analysis conducted to determine the preferred option among two or more options, such as the number of engines, based on a figure of merit, such as cost, weight, or reliability. Trade-offs can be either top level or subsystem level. At the top level, for example, the trade-off between payload and range will be paramount. Chapter 8 will discuss this trade-off and many other top-level trade-offs. Chapter 9 describes subsystem trade-offs, for example, the trade-off between electrical and pneumatic de-icing. There are hundreds of other subsystem-level trade-offs.

The importance of the trade-off study is that it is the key step which allows the designer to find the best solution for both the aircraft and its subsystems. Simply identifying a trade-off indicates that the previous design was, perhaps, not the best design for today's environment and the available technology.

Top-level trade-offs will be complete before the SDR so that a top-level design concept can be presented which meets the top-level requirements. Subsystem-level trade-offs are complete before the PDR, at which all requirements are complete. Hence, trade-offs should be conducted top-down. That is, the top-level trade-offs should be conducted before the subsystem-level trade-offs.

An important part of trade-offs is risk as discussed in Chapter 15. When the designer is considering two solutions, the performance, schedule, and cost risks should be considered for each solution. Risk assessment is particularly important when considering new aircraft technologies, discussed in Section 7.6.

7.4   Quality Function Deployment (QFD)

Quality function deployment (QFD) is a process which significantly improves the ability to capture, prioritize, and assess customer requirements using a team approach, and to transform these requirements into concepts. The strength of QFD is its ability to prioritize and weigh customer needs and to differentiate the value among different options. QFD has been shown to achieve satisfaction with both the customer and the manufacturer. Although a major function of QFD is to capture requirements, it is discussed here because it is most useful as a tool for establishing an initial concept.

QFD transforms customer needs into design characteristics and prioritizes the needs based on customer importance. These characteristics and priorities are translated into component characteristics. QFD can examine many factors, including functionality, reliability, cost, technology, and non-quantifiable aspects, such as aesthetic appeal. QFD can be applied at any level of DAC or customer operation, including design, development, manufacturing, training, operation, maintenance, and service.

In summary, some of QFD's strengths are:

1. Its ability to evaluate both quantitative and qualitative needs. It is this ability which provides a bridge between qualitative and quantitative needs. For example, QFD can be used with real airline customers to establish preferences and trade-offs on human factors discussed in Section 5.5 or interior design discussed in Section 9.4, fields not normally subject to quantitative evaluation.
2. Its speed and ability to reach important conclusions quickly.
3. Its ability to integrate the opinions and contributions from many design disciplines, customers, suppliers, and decision makers.
4. Its ability to evaluate technical, cost, and schedule risk.
5. Its ability to check decisions against *common sense*.
6. Its ability to identify the drivers which led to any given decision.
7. Its ability to identify key trade-offs.
8. Its ability to identify synergistic solutions: that is, the ability to allocate multiple requirements to a single solution.
9. Its ability to prioritize needs.

It must be remembered, though, that QFD is not a scientific methodology. It is a tool based on reason whose purpose it is to arouse the intuition of the analyst and apply it to the creation of a system. If the results agree with the intuition, then it can be said that the final solution is probably a good one. If not, the analyst should take advantage of the transparency of the methodology and determine how to improve the system.

The primary tool of QFD is the House of Quality. An example of a House of Quality is shown below:

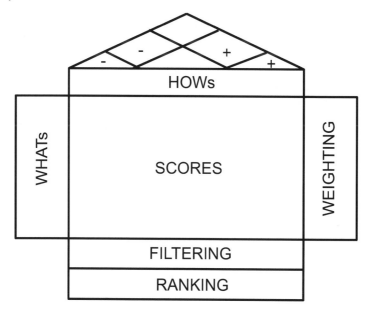

Figure 7.2   The House of Quality

In the House of Quality of Figure 7.2:

1. WHATs are the customer needs or requirements. The WHATs are collected in a group environment from the customer. WHATs can be top-level requirements, such as range, payload, or dispatch reliability; or special customer options, such as passenger entertainment.
2. The establishment of customer needs is an especially critical aspect of the QFD process. This is because customers, or even engineers, may not have a clear understanding of the difference between a need and a solution. It is important at this stage to focus on needs and not solutions.
3. For example, does the customer want an extra door in the cargo area, or do they want improved ventilation? Improved ventilation is a valid need, but an extra door is just one of many possible solutions. The comparison of these solutions is part of the synthesis process. Improved ventilation is the proper need for the QFD process.
4. HOWs are candidate solutions to the WHATs. HOWs are developed by the systems engineer and approved by the customer.

5. A SCORE is a numerical value showing the importance of a particular HOW to contribute to a WHAT. SCOREs are obtained in a joint workshop environment with the customer. Non-linear scoring systems have been found to be useful in identifying the desirable solutions. For example, a scoring system might be as follows: 0 (not important), 1 (slightly important), 3 (moderately important), and 9 (very important).

6. A WEIGHTING is a numerical value reflecting the importance of the WHATs to the customer. WEIGHTINGS are solely determined by the customer.

7. FILTERING is a numerical value designed to account for such factors as cost and schedule risk.

8. The RANKING is a numerical value determined by multiplying the WEIGHTINGS with the SCORES, summing the product for each HOW, and adding the FILTERING value.

The intersections of the roof of the house reflect either synergisms (+) or trade-offs (-) between the HOWs. For example, any customer option which results in an increase in weight will also result in a decrease in the range of the aircraft. When a final concept is selected (a group of HOWs), both the RANKING and the intersections are considered. Figure 7.3 shows a House of Quality with some typical WHATs and HOWs normally encountered in top-level aircraft analysis.

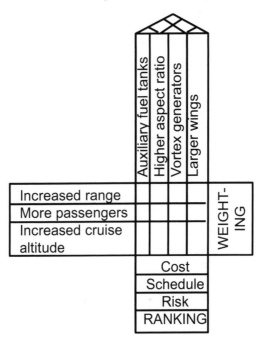

Figure 7.3   The House of Quality with typical WHATs and HOWs

7.5 Safety Features

Through the Preliminary System Safety Analysis (PSSA) discussed in Section 10.2 and the Common Cause Analysis (CCA) also discussed in Section 10.2, the certification process requires that safety become an integral part of the synthesis process. For example, the certification plan requires the description of any unique or novel feature of the design which may be a safety hazard. Other safety-related steps are described in Section 10.2.

7.6 Introduction of New Technologies

New technologies in the SE process

The introduction of new technologies is an integral part of the synthesis process. Section 2.4 presented a list of typical new technologies which may be considered. The question is then how the introduction is accomplished within the SE methodology. First, it can be done as part of the initial concept formulation described above. That is, the initial concept will have as an integral feature certain technological features which are prejudged to meet the top-level requirements and have been shown in the initial sizing as discussed in Section 8.2 to be effective both from a weight and DOC standpoint. Examples might be composite structures or fly-by-wire (FBW).

However, the crucial step for the evaluation of new technologies is during the trade-offs to be performed following the SDR. In SE the process of trade-offs leading to a synthesized concept is called *system analysis*. It is here where all aspects of a new technology need to be evaluated. These aspects include the ability to meet performance requirements, cost, weight, and risk.

Evaluation of new technologies

Mackey (1996) has outlined an eight-step strategy for technology management. The first step is to evaluate the requirements for a technology assessment process. These requirements include a thorough evaluation of the mission of the system (the aircraft) and an evaluation of return on investments and benefits of introducing new technologies. The next step consists of evaluating the readiness levels of candidate technologies. Table 7.1 shows the standard readiness levels adapted from NASA.

The next step is to make sure that customers, aircraft manufactures, for example, have direct contact with the universities and other institutions who are developing the new technologies. The fourth step is to gain the commitment of project management. With this commitment comes the necessary funding for the new technologies. The fifth and key step is a process called technology evaluation and adaptation methodology (TEAM) as developed by Loral (1994). TEAM performs a complete mapping of the mission of the system with the needed technologies.

Table 7.1   Technology readiness levels

Level	Description
1	Basic principles observed and reported
2	Technology concept and/or application formulated
3	Analytical and experimental critical function and/or characteristic proof of concept demonstrated
4	Component and/or breadboard validated in laboratory environment
5	Component and/or breadboard validated in relevant environment (ground or flight)
6	System or subsystem model or prototype demonstrated in a simulated environment (ground or flight)
7	System demonstrated in flight
8	Actual system completed and flight qualified through test and demonstration (ground or in flight)
9	Actual system flight proven

Source: Adapted from NASA (2012).

It also conducts an assessment of potential payoffs and risks on a technology roadmap. Finally, it develops a plan which documents schedules, funding, and conclusions. The sixth step is the use of a prototype or pilot process. Various types of prototypes can demonstrate requirements, integrate and test functionality, determine system feasibility, create simulated environments, and demonstrate operational concepts. The seventh step is to implement new technology on selected projects. The final step is to summarize and document lessons learned. In this way the steps already taken do not need to be repeated. In summary, in the aircraft industry the introduction of new technology is not just a matter of scientific interest or even of economic profitability. Rather it is a matter of survival.

7.7   Preliminary Design

The last real step in the synthesis process is preliminary design. The name may be somewhat misleading because the design is, in fact, final. The only aspect of the design not done is the detail drawings. At the PDR described in Section 12.4 all the trade-offs identified at the SDR described in Section 12.4 will have been completed. These activities, such as trade-offs, leading to a synthesized design are called *system analysis*. All the functions and requirements will have been completed to the lowest level of the aircraft hierarchy. And finally, the design will be complete to that level with weights and dimensions. These design characteristics are the *design requirements*, which are the product of the synthesis process.

8

Top-Level Synthesis

For as the aircraft is one system, and hath many subsystems, and all the subsystems of the aircraft, being many, are one aircraft …

Paraphrased from I Corinthians 12: 12–26, *Holy Bible*, King James (Authorized) Version

This chapter will focus on the creation of an aircraft system at its very highest level. It is not the purpose of this chapter to present the step-by-step procedure for the sizing of an aircraft. For the aircraft, these steps are well-known, for example, in Corning (1977). Rather it is to show how these steps fit into the SE process so that the parameters can be understood in terms of the basic SE categories, namely, performance requirements, constraints, and trade-off parameters. We will also show how these parameters logically flow down to the lower levels of the aircraft hierarchy. In this chapter we will look at how performance requirements and constraints result in both a synthesized aircraft system and a synthesized aircraft at the top level. Of course, the determination of the requirements at all levels will require the implementation of the principle of *holism* discussed in Section 4.3 and Chapter 7.

The creation and building of complex systems is often called *systems architecting*, as defined by Rechtin (1991). Systems architecting goes beyond technical requirements to focus on such concepts as customer satisfaction. This chapter will concentrate on the creation of an aircraft system from verifiable top-level technical and economic requirements.

8.1 The Aircraft System

We have learned in Section 2.3 that the real top level is the *aircraft system*, of which the aircraft is only one of element out of five. These elements are the aircraft, the training equipment, the support equipment, facilities, and personnel. Hence, when performing top-level synthesis, it is necessary to synthesize the aircraft system, not just the aircraft. This principle is particularly important for the high-speed civil transport (HSCT) for which radical changes in cargo or passenger loading or maintenance may be required. Following are a few example considerations for specifying the entire aircraft system; although many of these items will affect the design of the aircraft itself, they may also affect the design of the four other aircraft system elements:

1. *Cargo characteristics* This category includes weight, volume, count, and any special characteristics, such as live animals, explosives, and toxic materials.

2. *Actual origins and destinations* This item includes particular route capabilities, abilities to operate in severe weather, or at airports with particular characteristics affecting the take-off and landing approaches.

3. *Airport characteristics* These items include ramp, taxiway, and runway constraints; airport terminal constraints on aircraft dimensions and servicing locations; passenger access equipment; servicing equipment sizes, types, rates, and interfaces; maintenance facilities; lighting; personnel facilities; airline operations facilities; and any special rules.

4. *Configuration change-over times* This item applies to systems for which the aircraft configuration is required to change between flights.

5. *Utilization rate* This requirement measures how many hours per week is the aircraft required to fly, and how much support this requirement implies.

6. *Reliabilities, both dispatch and economic* These reliabilities drive both aircraft and support requirements.

7. *Turnaround time* This requirement drives aircraft cargo and passenger loading characteristics as well as maintenance and servicing features. Adams (1996) shows that on-board data loading can result in intolerably long turnaround times. A shop data loading system is proposed.

8. *Passenger service requirements* These requirements include food, lavatories, and airport passenger services, such as transportation.

9. *Growth capability* This requirement influences the need for commonality in all five elements.

10. *Autonomy* The requirement for aircraft starting and servicing autonomy results from the constraint to utilize less ground equipment.

11. *People-related requirements* These requirements cover all aspects of the system: for example, number of people, types, quality levels, cross-training requirements, certification needed. Included are dispatch operations, flight crews, cabin crews, servicing, maintenance, overhaul, training, and engineering. People-related requirements should be included in all of the other considerations in this list.

12. *Consumables* These items are included in all system elements: fuel, lubricants, sealants, coatings, life-limited parts, non-repairable items, interior (for example, seat covers, towels, and carpets), and perishable tools.

13. *Operational requirements* These requirements include air traffic control (ATC) compatibility, traffic at departures and destinations, communications, navigation, climb rates, initial cruise altitude (ICA), taxi rules, in-flight and ground aircraft and engine wakes, approach speeds, and cruise Mach number.

14. *Exterior noise* This requirement includes effects on maintenance, ground operations, and flight profiles.

15. *Regulatory environmental requirements* These include limits on nitrous oxide (NOX), unburned hydrocarbons, heavy metals, paint, coatings and

sealants, asbestos, soot and other particulates, fuel and other fluids dropped on ramp, discarded waste, and hazardous waste.

16. *Particular customer requirements* These requirements include any mandated solutions and all functional and physical interfaces. Also included are customer methods of transferring maintenance and dispatch data. Also included are customer constraints on supplier selection.

17. *Costs* In addition to the direct operating costs (DOC) discussed below and other indirect costs discussed in Section 8.6, these include all other segment costs. Examples are airport fees driven by exceeding noise and environmental limits, people costs associated with Item 11, spares cost, facilities cost, service costs, and training costs.

An example of an aircraft system-level function requiring synthesis beyond the aircraft level is Provide Category III (all weather) Landing Capability, an operational (Item 13 above) requirement. This function requires synthesis and qualification of the aircraft, the pilot, the maintenance system (support element), and the airport. These elements include the people associated with the elements.

8.2 Top-Level Aircraft Sizing

Top-level sizing is the heart of the creative process in aircraft design. It is also the beginning of the synthesis phase in the SE process as we saw in Figure 1.1. The first product of top-level sizing is an *initial concept*. This concept is the result of the top-level functions developed according to the principles of Chapter 3 and the requirements developed in accordance with the principles of Chapter 4. Most of the requirements considered in this step result from the Perform Air Transport Mission function of Figure 3.2. However, as we have seen, all functions should be considered. This initial concept is the subject of the system design review (SDR) discussed in Section 12.4. It is therefore the basis for further trade-offs and optimization.

Wing sizing

During the Perform Transport Mission function discussed in Section 3.2 it is necessary to size the wing by balancing three conditions: take-off, cruise, and landing. The following performance requirements apply:

1. Number of passengers.
2. Weight of cargo.
3. Range.
4. Cruise Mach number.

It is the basic mission of the aircraft to carry a specific number of passengers and cargo a given distance in a given time. Hence, these parameters are performance requirements. The following constraints also are required to size the wing:

1. Field length.
2. ICA.
3. Atmospheric density, pressure, and temperature at each of the three conditions.
4. Approach speed.

The field length is set by the route profile of the customer. Approach speed and ICA can also be traded off against payload weight.

Wing sizing trade-offs

The trade-offs for wing sizing are extensive as listed by Corning, who also provides the detailed equations to conduct these trade-offs.

Supercritical vs. critical wing	Specific fuel consumption (SFC)
Aspect ratio, AR	Engine type
Sweepback angle, $\Lambda_{C/4}$	Fuel/take-off weight ratio, W_f/W_{to}
Divergence Mach number, M_{DIV}	Take-off wing loading, $(W/S)_{TO}$
Average thickness/chord ratio, $(t/c)_{AVE}$	Initial Cruise wing loading, $(W/S)_{IC}$
Maximum lift coefficient, $C_{L_{max}}$ (no flap extension)	Initial cruise lift coefficient, $C_{L_{IC}}$
Maximum lift coefficient, $C_{L_{max}}$ (landing)	

In order to find the optimum combination of all these parameters it is necessary to balance the requirements in the three conditions described above, namely, take-off, cruise, and landing. Even then, it will not be certain that the fuel the wings can carry will meet the range requirements. So it will be necessary to repeat these steps as the aircraft is defined.

Engine sizing

The engines are normally sized by the conditions at take-off. For some aircraft the engines are sized by the cruise or climb conditions. Therefore, it will be necessary to repeat the sizing steps for these conditions. The key performance requirement is thrust at take-off. However, it is necessary to derive the engine performance requirement from the results of trade-offs with other aircraft parameters. Other key constraints which are needed to size the engines are:

1. Take-off conditions:
   - Field length.

- Obstacle height at end of runway.
- Density, pressure, and temperature at take-off.
2. From wing sizing:
   - Wing loading, W/S.
   - Maximum take-off lift coefficient, $CL_{t/o}$.
   - Number of engines.

From this information, it will be possible to determine the thrust loading W/T at take-off and the take-off speed, V_{to}. The engine thrust is sized by the thrust required at take-off. The basic parameter required at this point is thrust loading, W/T, at take-off. These basic performance requirements and constraints will provide both thrust loading and the take-off speed, V_{to}.

Take-off weight

The take-off weight is the fully loaded weight as described in Section 5.2. However, in the synthesis process it is not a constraint; it is a derived requirement. However, in the end the airport runway strength will set the maximum value of the take-off weight. So, for that purpose, it is a constraint. It is possible to determine the take-off weight from the following equation:

$$W_{to} = W_{structure} + W_{engines} + W_{fuel} + W_{payload} + W_{fixed\ equipment}$$

In addition, it is possible to determine the structure weight from:

$$W_{structure} = W_{wing} + W_{fuselage} + W_{landing\ gear} + W_{nacelle\ \&\ pylon} + W_{tail\ surfaces}$$

Each weight component is a function of the total take-off weight and several other parameters. Thus, we can solve for take-off weight. We will treat each one of these separately:

1. Wing weight (W_{wing}) is a function of the take-off weight (W_{to}), the aspect ratio, AR, the taper ratio, 1, the thickness ratio (t/c), and the sweepback angle, $L_{C/4}$.
2. The fuselage weight ($W_{fuselage}$) is a function of the number of passengers, the number abreast, and the number of aisles.
3. The landing gear weight ($W_{landing\ gear}$) is normally a fixed fraction of the total take-off weight.
4. The nacelle and pylon weight ($W_{nacelle\ \&\ pylon}$) is proportional to the thrust and hence is a function of the take-off weight and thrust loading.
5. The tail section weight (W_{TS}) is proportional to the wing weight and hence to the take-off weight.
6. Like the nacelles and pylons, the engine weight can be assumed to be proportional to thrust.

7. The fuel weight (W_{fuel}) can be assumed to be a fixed fraction of the take-off weight.
8. The payload weight ($W_{payload}$) is the total assumed weight of the passengers and cargo from the performance requirements as discussed in Section 8.2.
9. The fixed equipment weight ($W_{fixed\ equipment}$) includes all subsystems including electrical, environmental control, and so forth. Driving factors are the number of passengers, number of engines, and total take-off weight.

When these factors are added, it will be possible to solve for the take-off weight. Yet, even now, we have not finished our trade-offs. We do not know whether the take-off weight exceeds runway limits or whether the aircraft can carry the required fuel weight.

Drag trade-offs

The drag coefficient of the aircraft is determined from the basic equation:

$$C_D = C_{D0} + C_L^2/\prod ARe + \Delta C_{DC}$$

Each term is affected by a different aspect of the aircraft configuration. C_{D0} is the lift-independent, drag, which is a function of the total aircraft configuration. It contains both form and skin friction drag. The second term, $C_L^2/\prod ARe$, is the induced drag. This factor depends on the total lift coefficient, C_L, which was determined in the wing sizing step above, as that necessary to maintain level flight. The aspect ratio, AR, was also a trade-off parameter in the wing sizing. The lift efficiency, e, and the compressibility drag coefficient, ΔC_{DC}, are also design-dependent factors, and therefore, *derived requirements*, as is the total drag coefficient C_D.

Lift-drag ratio

As we will see later, a key parameter in the determination of aircraft range is the lift-drag ratio, L/D, which is now determined by:

$$L/D = C_L/C_D$$

Climb requirements

With the initial engine sized above, the estimated aerodynamics, and the ICA, it is now possible to estimate the amount of fuel required to climb to the cruise altitude and the range to climb, R_{cl}.

Cruise range

The range during cruise (assumed to include descent) can be estimated by the famous Breguet range equation:

$$R_{cr} = (V/SFC)(L/D)\log_{10}(W_1/W_0)$$

where V is the speed, W_0 is the initial cruise weight, and W_1 is $W_0 - W_{fuel}$. Thus, the total range is:

$$R = R_{cl} + R_{cr}$$

It is possible to vary the range by varying the fuel fraction in the wing sizing step. Hence, an estimate of the total aircraft sizing results.

Top-level derived requirements and allocation management

What we learned in the above steps was not how to size an aircraft, but rather that the sizing process is an initial requirements process in which a set of top-level performance parameters and constraints are established and top-level trade-offs are conducted. These steps as described were only very approximate.

Another product of this process is a large array of *derived requirements*, that is, requirements which are dependent on a solution. Thrust is an example. This requirement is then used to size the propulsion subsystem of the aircraft. All of the other requirements will be used in the flow down of requirements to the aircraft subsystems.

Another aspect of this sizing process is that it gives the SE manager a tool to *manage* the allocated parameters, such as weight. We saw in Section 4.7 how requirements could be allocated among subsystems and then adjusted as trade-offs showed more optimal allocations. We will see in Section 12.3 how the SE manager uses these allocations to manage the design. The sizing exercise above provides us with the initial allocation values.

8.3   Other Top-Level Requirements

Remember that we defined many top-level aircraft functions described in Section 3.3. The sizing analysis above will satisfy some of the key functions, for example, Generate Aero Forces also in Section 3.3 and Provide Thrust, also in Section 3.3. However, many of the other functions will result in top-level (not derived) requirements which may have a significant impact on the aircraft. These requirements will depend on the particular needs of the customer.

The Provide Environmental Control function from Section 3.3 is a normal top-level function which may, or may not, reflect any special customer needs. The basic requirements which are *allocated* to this function are the temperature, pressure, humidity, and air quality throughout the aircraft. The importance of listing these as top-level requirements is that many subsystems (for example, ECS, airframe, and propulsion) will contribute to their achievement and that trade-offs may have to be made to determine the optimum allocation of requirements among these subsystems.

The requirements allocated to the Provide Passenger and Crew Accommodations function from Section 3.3 may vary widely among customers. For example, a customer may need special seating arrangements or lavatory or galley configurations. Once again, many subsystems may be involved in satisfying these requirements, and trade-offs may be required.

8.4   System Architecture

Before we can flow down any requirements to the subsystems, we need a *system architecture*, that is, the hierarchy of aircraft elements. We have already formulated a typical aircraft system architecture shown in Figure 2.1. This is not the only possible system architecture. Like the initially sized aircraft above, the system architecture is also subject to trade-off. For example, it is entirely possible to design an aircraft without one or more of the elements shown in Figure 2.1. One could design an aircraft, for example, without hydraulic power, using only electrical power. If such an aircraft could ever be optimum is a matter subject to trade-off. Figure 2.1 shows, of course, only the top-level system architecture. Each element in that architecture would have its own subelements, also subject to trade-off.

A key factor in the development of the system architecture is system safety. As we will see in Section 10.2 the preliminary system safety analysis (PSSA) is used to help develop the architecture.

8.5   Top-Level Constraints

In addition to the performance requirements listed above, and the associated derived requirement we already developed from the initial sizing, we should now address the multitude of top-level constraints.

Airport compatibility will introduce many constraints: For example, the weight limit of the runway and the width of the gates. Support capability at airports will determine the support design philosophy. For example, should the aircraft have autonomous support capability? That is, is it required to land, restart, and take off without any ground support? All of this is determined from the route structure the aircraft is being designed for.

Two key top-level constraints are dispatch reliability and maintenance cost per 1,000 flight hours (MN\$/1000FH). As we have seen in Section 4.7, we can allocate these requirements to the subordinate elements. Other top-level constraints are cabin and exterior noise and emissions, which can also be allocated.

Design constraints, environments, and regulatory requirements, as discussed in Chapter 5, can be said to be constraints because they apply to the whole aircraft.

The most important constraint is the cost, normally characterized as DOC. Because of its importance, we will save cost for a fuller discussion in the next section.

8.6   Economic Constraints

In the SE process, cost may be considered a design constraint as valid as any other technical constraint, such as weight or physical limitations. Such a constraint may be incorporated directly into any specification.

In military practice, design-to-cost is normally treated as a trade-off activity to be conducted parallel to the design activity. In contrast, the demands of the competitive market require that cost constraints be treated more directly because market development is based on a given selling price.

Not only can costs be imposed on a system, an aircraft for example, but the cost constraints can be flowed down to the system segments and subsystems just as any other technical parameter can.

To accomplish this flow down, we can divide cost into three categories: non-recurring (development) costs, recurring (unit) costs, and DOC. Within DOC, major categories can be identified and used as design constraints. Although the components of DOC vary slightly from source to source, the major ones cited by Martínez-Val (1994) and Aerospace Engineering (1994) are shown in Table 8.1. All three major categories are, of course, of interest to both the developer and the customer since they affect both system purchase cost and operating cost, and hence profitability of the customer and, therefore, sales by the developer.

Although most aircraft economic analysis is focused on DOC and recurring costs, non-recurring cost is also a major driver in aircraft requirements. Non-recurring cost is the factor which leads to the need for derivative aircraft discussed in Section 2.2. By basing the design on a previous aircraft, the aircraft manufacturer can save a considerable amount of money in development costs. This savings goes beyond the development of the aircraft itself. Derivative aircraft result in savings in tooling and jigs used in the manufacturing process.

Direct Operating Cost (DOC) requirements

Figure 8.1 shows how the SE process can be used to allocate requirements. First, the aircraft DOC goal is allocated to the various aircraft segments in accordance with the items in the following list:

- navigation fees
- landing fees

- ground handling
- crew (cabin, cockpit)
- ownership (depreciation and interest)
- maintenance (fuel and airframe)
- fuel and oil

Figure 8.2 shows how DOC is used to select the design point for a new aircraft. Two types of DOC are important: DOC per seat-mile, and DOC per trip. Design points in the lower left-hand corner are deemed to be economically viable while those in the upper right-hand corner are deemed not to be viable. These two regions are the result of an economic analysis which can be conducted either by the airline or the aircraft manufacturer.

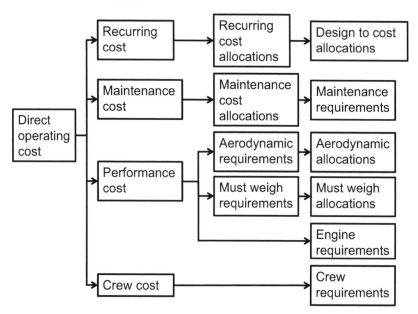

Figure 8.1   Allocation of direction operational cost in design process

Initial Direct Operating Cost (DOC) estimate

Before we can allocate DOC to the various aircraft subsystems, we have to have an initial estimate of the DOC breakdown. If this procedure sounds familiar, it is exactly the same as for the weight breakdown which we saw earlier in this chapter. DOC (in dollars per ton-mile) is comprised of three main factors: flight operations, direct maintenance, and depreciation. This initial DOC estimate is a direct result of the initial sizing discussed earlier in this chapter. The DOC will depend on the top-level requirements and the parameters estimated in the initial sizing, for example, take-off weight and number of engines.

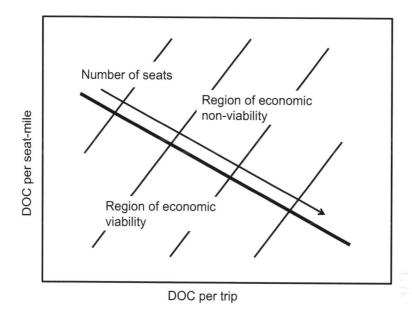

Figure 8.2   DOC design regimes

Flight operations
Two main factors need to be calculated first to estimate the flight operations component of DOC, namely, block speed, V_b, and block fuel, F_b. Block speed is simply the total range divided by the time from gate to gate. Likewise, block fuel is the total fuel used from gate to gate.

Flight crew
Flight crew costs can be found by available models with cost primarily as a function of aircraft size.

Fuel and oil
Fuel costs were already determined from the block fuel used. Similarly oil costs can be determined from the block time and the number of engines.

Hull insurance
The insurance cost is based on the aircraft unit cost and the utilization factor.

Direct maintenance
Direct maintenance cost is comprised of the following three components:

Labor (excluding engines)   This is the primary parameter in labor cost models of the aircraft. This factor excludes the labor cost associated with the engines.

Labor (engines)   The labor cost associated with the engines can be estimated knowing the size (thrust) and number of engines.

Material   Maintenance material cost is most strong correlated with the total cost of the aircraft.

Depreciation cost

Depreciation is the primary DOC component which accounts for the unit (recurring) cost of the aircraft, although the hull insurance component is also dependent on the unit cost. The depreciation is dependent on the unit cost, the depreciation period, and the utilization factor, U, of the aircraft. Normally, the depreciation of the aircraft and depreciation of the engines are calculated separately and added together.

Allocation of cost to system segments

Table 8.1 shows how a specification format can allocate the three cost categories to the major aircraft segments. In the military world, specifications are written to document and allocate requirements at all levels of a system, whether or not one contractor makes that system.

Table 8.1   Allocation of cost constraints

Aircraft segment	Recurring costs	Non-recurring costs	DOC
Airframe	X	X	X
Avionics	X	X	X
Environmental	X	X	X
Mechanical	X	X	X
Electrical	X	X	X
Interiors	X	X	X
Propulsion	X	X	X
Support equipment	X	X	X
Assembly	X	X	N/A
Development	N/A	X	N/A
Production support	X	N/A	N/A
Total cast	X	X	X

Aircraft cost allocation

We saw above that the recurring cost of the aircraft is inherent in the depreciation and hull insurance components of DOC. With the initial sizing concept we can estimate the total cost of the aircraft using standard industry cost estimating techniques. These cost estimates provide a cost allocation for each subsystem. These cost allocations become the basis for the design-to-cost processes to be used on each subsystem. These cost allocations, also, can become technical performance measures (TPMs) as a subject of program management described in Section 12.7.

Performance allocation

Performance cost allocation must be done in a different manner since it cannot be allocated directly to all segments. There are three primary factors: aerodynamic performance, weight, and engine performance. Each of these factors can, then, be allocated to the appropriate segments. In addition, many components use power and have volume which affects drag. All three must act in concert to maintain the performance cost to a given level. The beginning of this chapter describes how to make an initial estimate of these performance parameters. Each of these parameters becomes a derived requirement which must be managed in order to meet the total estimated DOC.

Aerodynamic parameters

The initial sizing set the aerodynamic parameters (drag and lift coefficients) at the aircraft level to meet the performance and cost goals. The components of the total aerodynamic coefficients can then be established and allocated to the major external airframe components (wing, fuselage, empennage).

Weight allocation

Weight, in contrast, can be allocated directly. Once the total weight requirement of the aircraft is established, the weight of each segment and subsystem can be determined by allocation.

Engine performance

Next, the engine performance requirements established by the initial sizing become the basis for the engine performance contribution to DOC.

Maintenance cost allocation

The initial sizing discussed above provided an initial estimate of the total maintenance cost in terms of maintenance dollars per one thousand flight hours (MN\$/1000FH). In practice maintenance cost per flight hour varies significantly among subsystems as shown by historical data. This historical data can be used to establish maintenance cost allocations to the subsystems. Of course, the total maintenance cost cannot exceed the value estimated in the initial sizing.

Flight crew requirements

Finally, the crew requirements can be established from the allocated crew cost. Flight crew requirements are a type of people-related requirements discussed in Section 8.1, Item 11. The requirements include the number and skill types for flight deck crew and flight attendants. These requirements are normally fixed by regulation and do not often figure into aircraft optimization. However, cockpit requirements can be established to minimize the flight crew training.

Indirect cost factors

Like DOCs, indirect costs can also be used to establish design requirements. Major contributors to indirect costs are dispatch and operational reliabilities. For dispatch reliability, for example, the airline will incur direct costs, in terms of lost ticket sales and other factors, for each aircraft which fails to be dispatched on schedule. The impact of dispatch failures may vary greatly among airlines depending on the airline operating characteristics. Other indirect costs include ferry flights for repair and replacement and direct passenger costs for missed connections. Like weight, the dispatch reliability can be allocated to the aircraft segments and to their subsystems. Hence, the standard SE practice of requirements allocation can be used directly to flow down performance requirements to all system segments, subsystems, and components to meet operational direct and indirect cost constraints.

Manufacturer's indirect cost savings

SE is expected to result in other cost savings to the manufacturer. However, these savings cannot easily be converted into design requirements. Many of these savings can be considered as SE *metrics*, that is, a measure of the effectiveness of the process, providing the contribution of SE can be isolated. A discussion of metrics, however, is beyond the scope of this book.

Honour, for example, shows the cost savings in project management from using automated SE tools (Honour, 1994). A successful metric is the cost reduction from fewer redesigns and after drawing release, both in engineering and manufacturing. Others are reduced warranty claims and improved sales. Although SE may have a significant impact on sales, the effect would be virtually impossible to isolate.

Airline customer's indirect cost savings

Similarly SE is expected to result in many other cost savings to the airline customers. These savings also may be impractical to convert into design requirements unless cost models exist which permit this conversion. Indirect airline savings include: savings in reservations and ticket sales, saving in advertising and publicity, reduced maintenance and depreciation of non-flight items, reduced general and administrative (G&A) costs, and reduced passenger services (transportation and hotels). Of these, only those which can be linked to dispatch reliability are usually candidates for design-related costs.

8.7 Top-Level Trade-Offs

In addition to the sizing trade-offs discussed earlier in this chapter, certain other trade-offs count as top-level because they involve trade-offs between (or among) segments. Some typical top-level trade-offs are shown in Table 8.2. Chapter 9 will discuss these on a subsystem-by-subsystem basis.

Table 8.2   Top-level trade-offs

Trade-offs	Subsystems
Mechanical vs. electrical controls	Mechanical, electrical
Electrical vs. hydraulic power	Electrical, hydraulics
Bleed-air vs. self-contained air supply	Pneumatics, propulsion
Noise suppression	Fuselage, interiors, propulsion, electrical (active suppression)
Conventional vs. propulsion control	Propulsion, mechanical, electrical, airframe

9

Subsystem Synthesis

And the avionics cannot say unto the pilot, I have no need of thee, nor again the empennage to the wings, I have no need of you.

Paraphrased from I Corinthians 12: 12–26, *Holy Bible*, King James (Authorized) Version

This chapter will show the primary performance and constraint requirements for each of the major subsystems. It will enumerate and describe the principal subsystem-level requirements parameters and how they are allocated to subsystem hardware and software in the SE methodology.

As we have noted before, subsystems are normally called systems in the aircraft industry. However, in SE terminology *subsystem* better describes where these elements fit within the aircraft hierarchy and methodology.

Virtually all subsystem requirements are derived requirements. That is, they depend on solutions for their values. Therefore, we cannot, in this chapter, say which requirements are allocated to which subsystems. We can only describe typical allocations. Neither can we say, in all cases, what the exact, quantitative requirements are. We can only determine what the functions are that should be converted into quantitative requirements and subsequently allocated to hardware or software.

In addition, it will be remembered that a basic SE philosophy calls for the requirements for the subsystems to be *holistically* determined as described in Section 4.1, and the conflicting requirements must be resolved using the methods of requirements trade-offs described in Section 4.8.

As we saw before in Section 8.4, the flow down of requirements from the top level is dependent on the aircraft architecture selected at that level. We are using the architecture described in Figure 2.1, keeping in mind that other architectures are possible and that top-level trade-offs will be necessary to determine whether the architecture of Figure 2.1 is, indeed, the best architecture.

A key aspect of synthesis is the trade-off. We cannot say exactly what trade-offs need to be made; we can only describe *typical* trade-offs. Subsystems will be involved in trade-offs at two levels: the aircraft and the subsystem level. Although Chapter 8 dealt with aircraft-level synthesis, we will discuss here specific instances in which subsystems may be involved in trade-offs of two or more subsystems. The importance of these inter-subsystem trade-offs is that they should be addressed very early in aircraft development, that is, before solutions are *assumed* based on the experience of a single design discipline.

Implicit in the synthesis of every subsystem discussed below is the principle that quantitative requirements should be developed from each function in accordance with the principles of Chapter 4, and that these requirements should drive the subsystem design. We will continue to capitalize the names of functions in accordance with the conventions of Chapter 3.

Since the segments below are not true subsystems, but rather collections of equipment, the correlation between these segments and the functions of Section 3.3 is only approximate.

9.1   Environmental Segment

Most functions provided by the Environmental Segment emanate from the Provide Environmental Control function in Figure 3.6.

Air conditioning (ATA 21)

The air conditioning subsystem has a number of key functions: Provide Temperature Control of Air Supply, Provide Pressure Control of Air Supply, Provide Distribution of Conditioned Air, Provide Air Filtration, and Provide Ventilation. It is theoretically possible to control the humidity of air in the aircraft. However, conditions rarely warrant humidity control. These functions are subordinate to the Provide Environmental Control function in Chapter 3.

While the traditional method of temperature control is to mix the warm air from the pneumatic subsystem with refrigerated air, this trade-off should be revisited for future designs. Alternatives include an independently controlled autonomous air supply with, perhaps, a ram air supply. A top-level trade-off would then be required to determine whether the ram air drag penalty on the aircraft would be too detrimental.

The design of the air conditioning system is highly dependent on the heat loads. The driving external environment is normally the hot-day ground condition. Internal heat loads, such as from electronic equipment, also figure in the sizing.

The Provide Distribution of Conditioned Air function provides the opportunity to minimize the total subsystem weight by optimally placing the air conditioning elements throughout the aircraft. The weight of cables and other components may be decisive factors in the placement trade-off.

In addition, of particular importance are both the *functional* and *physical* interfaces discussed in Chapter 6 the distribution subsystem has with other subsystems. For example, the galley and environmental engineers should be in complete agreement on the flow rate and characteristics of the air to be delivered to the galleys.

While the environmental control engineer has control of defining the air conditioning equipment, the flow of air throughout the aircraft can be considered a top-level issue because the air passes through, under, in, and around components not under the direct control of the environmental control subsystem (ECS). These components include bag racks, equipment racks, lavatories, galleys, tunnels, and cargo areas. This top-level consideration of functions normally considered the purview of subsystem organizations is a key value-added aspect of SE.

Cabin pressure (ATA 21)

Typically, the cabin pressure subsystem does not *provide* the cabin pressure. That is done by the air conditioning subsystem. The main functions of this subsystem are Monitor Cabin Pressure and Control Cabin Pressure Outflow. The sizing of the outflow valves will be dependent on the cabin pressure and the expected rate of out-flowing air to maintain the cabin pressure. The requirements for cabin pressure vary with altitude and are strictly controlled to prevent an excessive pressure change on the pressure shell of the aircraft.

Another function of the cabin pressure subsystem is Control Air Flow in the cabin. It is a constraint of aircraft design that air should flow laterally and not longitudinally in the cabin. Optimal placement of the outflow valves is required to meet this constraint.

Ice and rain protection (ATA 30)

The ice and rain protection subsystem is not a single collection of elements, but rather separate elements each with its own synthesis solution. The functions are Provide Anti-Icing, Provide De-Icing, and Provide Rain Protection.

De-icing is normally performed, for example, with hot air from the power plant delivered by the pneumatic subsystem. However, as we saw in Section 9.1, air can be provided from other sources. Anti-icing of small structural elements, such as strakes, can be done with a low current electrical heater. All of these solutions involve considerable trade-offs among the ice and rain protection, the electrical subsystem, the pneumatic subsystem, the power plant subsystem, and the fuel subsystem, with the figures of merit being weight, reliability, and fuel consumption.

Rain protection is normally accomplished with wipers and fluids.

Oxygen (ATA 35)

There are three oxygen subsystem functions: Provide Crew Oxygen, Provide Passenger Oxygen, Provide Portable Oxygen. The FARs provide the primary requirements for these functions. The main subsystem trade-off is between chemical and gaseous oxygen supplies. The oxygen duration requirement is driven by the mission profile of the aircraft. That is, the longer it takes to descend to a safe altitude, the larger the oxygen supply should be. Longer durations usually lead to gaseous oxygen supplies.

Pneumatic (ATA 36)

The pneumatic subsystem is similar to the cabin pressure subsystem in that it does not *provide* pressure to anything. The pneumatic subsystem acts as a conduit between the power plant and the air conditioning and ice and rain protection subsystems, for example. The pneumatic subsystem also provides air for engine starting, cargo heating, and water pressurization. Hence, the main function of the pneumatic subsystem is Maintain Pneumatic Pressure. Other functions are Provide Ozone Conversion of the exterior air and Provide Particle Filtration of recirculated air. However, as we saw in Section 9.1, all of these functions can be provided by alternative air supplies and are, hence, subject to top-level trade-offs.

Structural cooling

Conventional subsonic aircraft do not need nor do they have structural cooling subsystems. However, the high-speed civil transport (HSCT) may require cooling of its structural elements, primarily the wing, for high-speed flight. Trade-offs will be needed to determine the requirements and optimum technique for this cooling. The use of ducted fuel is one possible method for structural cooling.

9.2   Avionics Segment

Almost all functions provided by the Avionics Segment are subfunctions of the Navigate Aircraft function in Figure 3.6.

Auto flight (ATA 22)

The Navigation Aircraft and Commend and Control Aircraft functions are discussed in Section 3.3; these functions can be allocated either to the auto flight subsystem, to the flight crew, or to both. Factors associated with human flight control are discussed as part of the human factors analysis discussed in Section 5.5. The Provide Auto Flight function is subordinate to the Navigate Aircraft function is the equipment-related part of this function pertaining to the automatic control of the aircraft.

The Provide Auto Flight function uses air data, inertial navigation system (INS), and FMS as inputs and pitch and roll commands to the ailerons, elevators, rudder, and throttle. It provides the macroscopic commands for aircraft take-off, cruise, and landing and also the corrective commands, such as yaw damping, angle of attack correction, and Mach trim. It also performs speed and altitude correction by converting pressure altitude to actual altitude when the aircraft is below 10,000 ft.

Communications (ATA 26-10)

All communications functions are subfunctions of the Communicate Data/ Information function top-level function shown in Figure 3.6. The two principal subfunctions are Provide External Communications and Provide Internal Communications.

External communications

The requirements for external communications are driven by the range of communications, the need to transmit both voice and data, and the need to transmit free of atmospheric interference.

At shorter ranges, for example, nearer airports, very high frequency (VHF) systems have provided voice and data links. This medium is limited to line-of-sight communications and is not vulnerable to atmospheric interference.

For worldwide communications, high frequency (HF) links have provided non-line-of-sight communications, however, with the disadvantage of atmospheric interference. A more recent development in worldwide communications is SATCOM. Because of its accuracy and freedom from atmospheric interference and line-of-sight constraints, this medium is capable of replacing both VHF and HF media. Such a system will be essential for the HSCT.

The communications system is only responsible for transmitting voice and data. Process of this data should be accomplished by other systems, such as the ARINC (originally Aeronautical Radio, Incorporated now part of Rockwell Collins) communication addressing reporting system (ACARS) described in Section 9.2.

Another adjunct communications system is selective calling (SELCAL). The purpose of SELCAL is to send a coded signal from the ground through VHF, HF, or SATCOM to ring a chime in the flight deck indicating a desire for communications on the same frequency.

Another external communications system is the airborne telephone. This system operates on L-band on a line-of-sight path to ground stations. Messages are then relayed via ground and satellite to other ground stations throughout the world. SATCOM also has an embedded telephone system.

Because of the importance of external communications, much redundancy is required among external communications systems. Although most communications systems are embedded in the flight deck equipment, some aircraft systems require portable external communications as backup systems.

Static discharge wicks are located on various parts of the aircraft to dissipate static discharge which may cause external communications interference.

A final external communications system is the crash position indicator. An emergency locator transmitter (ELT) emits an emergency locator beacon (ELB) to determine the location of the aircraft.

Internal communications

The purposes of internal communications are for the flight crew and cabin crew to communicate with each other and to the passengers, and for service personnel to communicate with each other or with the pilots.

The passenger address system allows the flight and cabin crew to communicate to the passengers. This can be done either by audio (tape or voice) or by video (tape only). Because of the importance of this system in emergency situations, the passenger address system is powered by the backup battery power system described in Section 9.3.

This is another example of incorporating people into the requirements process. People, that is, flight and cabin crew and passengers, are a part of the system to which tasks can be allocated, for example, communicating emergency messages. They are also elements which interface with the system.

The services interphone system allows service personnel to communicate with each other or with the pilots. This function is part of the Provide Internal Communications discussed in Section 3.3 and is implemented by jacks located throughout the aircraft.

The flight interphone system allows the flight deck crew and the cabin crew to communicate with each other.

Other internal communication systems include the call system which allows passengers to signal flight attendants by means of a light. On-board megaphones provide backup communications to the passengers in emergency situations.

Audio mixing
A key requirement is for audio signals from all communication systems to be mixed and provided to the pilots. The system which does the mixing is the digital core avionics system (DCAS). DCAS integrates voice inputs from the headset microphone, the handset microphone, and the oxygen mask microphone, and provides the signals to the headsets and to the loud speakers.

Indicating and recording (ATA 31)

Instrument panel information
The indicating and recording subsystem provides the pilots with the critical information determined by the guidance and navigation subsystem discussed in Section 9.2 and other subsystems. Most requirements of the indicating and recording subsystem derive from the Communicate Data/Information top-level function discussed in Section 3.3. The instruments provide flight information regarding position, attitude, and heading.

The information provided by the instrument panel includes time (GMT) which is used as a time base for maintenance and for the flight recorders.

The layout of the instrument panel is strongly driven by human factors considerations discussed in Section 5.5. These considerations have led to the so-called *basic T* layout which, by convention, is used throughout the world, as shown in Figure 9.1.

The instrument panel is required to provide identical basic T information to both the pilot and co-pilot. In addition, a *comparator* determines the differences, if any, between the two readings. If there is a difference, the pilots should rely on the backup systems to determine which is more reliable.

Human factors analysis imposes other requirements to assure readability of the instruments, for example, parallax and glare avoidance, lighting levels, color coding, and font or needle sizes and shapes.

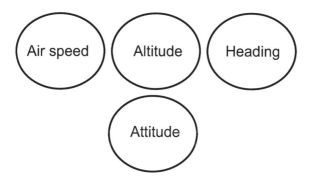

Figure 9.1   The basic T instrument panel layout

A design solution created by human factors is the so-called *dark cockpit* philosophy, which derives from the requirement to avoid distractions to the pilots. In the *dark cockpit* only the information absolutely needed is provided to the pilot.

Annunciation
Another key function of the indicating and recording subsystem is Provide Annunciation. Annunciation is any kind of visual or aural indication of aircraft status. Visual annunciations can take the form of lights or color displays on the instrument panel. Aural annunciations can take the form of many sounds, such as chimes, horns, or human voices. On modern aircraft a central warning system creates synthesized sounds. There are three levels of annunciation:

1. *Warning level*   Immediate action is required.
2. *Caution level*   Action is required but is not immediate.
3. *Advisory level*   An advisory annunciation indicates that an event has occurred.

Satchell (1993) provides a comprehensive description of monitoring and alerting systems for aircraft.

Data recording
The Record Data function is divided into three subfunctions: Record Cockpit Voice, Record Flight Data, and Record Maintenance Data. The first two are, of course, essential for accident investigations and are imposed by regulatory agencies. The maintenance data are collected to conduct trend analyses for failure prevention and to measure engine degradation, for example. The flight data and maintenance data are often recorded on the same recorder for economy.

Fault collection
A function is required to collect and record root cause information for faults, such as the failure of components. The system which does this is called the central fault

collection system. It provides sufficient information so that the cause of the faults can be diagnosed and corrective action can be taken.

Automatic data reporting

Another function of the indicating and reporting subsystem is to provide information to and from ground maintenance personnel during flight. This capability allows ground maintenance personnel to interrogate the aircraft and analyze maintenance data during flight. It also provides automatic reporting of maintenance data to the ground. The system is called the ARINC communication addressing reporting system (ACARS).

Navigation (ATA 34)

The navigation subsystem is an example of a subsystem derived almost entirely from a single top-level function, Navigate Aircraft shown in Figure 3.6. The requirements for the parameter associated with the named subfunctions vary according to (a) the degree of accuracy required, (b) the reference point of the information, and (c) the degree of redundancy required for safety.

The Determine Location of Aircraft subfunction can be divided into several categories. The first major category is dependent position, that is, dependent on ground stations. The Marker Beacon position, for example, gives position relative to the runway threshold. The VOR (VHF omni-directional radio) position gives bearing relative to north. The DME (distance measurement equipment) gives position relative to the VOR station. The combination of the two, bearing and relative position, gives absolute position.

The ILS (instrument landing system) gives position relative to airports. The second major category is independent on position, that is, independent of ground stations. This position can be determined by an inertial navigation system (INS). The rate of change of this position provides ground speed. Another, more recent, independent system is the global positioning system (GPS), which determines position from satellite information. Because of the high accuracy of the GPS, this system will probably be the preferred solution of the future.

Other critical pieces of position information are the positions relative to the ground and the prediction of the aircraft position relative to the terrain and to other aircraft. This information is used to provide warning of possible ground or aircraft impact described in Section 9.2. The system which provides prediction of aircraft collision is called the traffic collision avoidance system (TCAS). The system which predicts possible ground impact is called the ground proximity warning system (GPWS).

The basic solution for the Determine Heading of Aircraft function is the horizontal situation indicator (HSI), an inertial instrument. As a backup system, a magnetic whisky (that is, in alcohol) compass is used. The pitch and roll attitudes are determined by the inertial attitude direction indicator (ADI).

The navigation subsystem uses air data to satisfy a number of navigation functions. These include Determine Air Speed, Determine Angle of Attack and Slip, Determine Air Temperature, and Determine Impending Stall.

A major navigation function is Provide Flight Management. The purpose of the flight management function is to provide the correct aircraft commands to both the pilot and the auto pilot discussed in Section 9.2. The flight management system (FMS) is a computer which integrates the navigation data described above and provides this information to the Flight Director, another computer, which generates the commands. These commands are provided to the pilot through the instrument panel described in Section 9.2 and directly to the auto pilot. On some aircraft, pilot commands are provided to the aircraft directly from the Flight Director. The considerations involved in the trade-off between these two approaches are load feel characteristics, weight, reliability, cost maintenance, and training costs.

9.3 Electrical Segment

The primary top-level function in Figure 3.6 that the Electrical Segment supports is the Provide Power function.

Electrical power (ATA 24)

The primary electrical power functions are Provide AC Power and Provide DC Power. Necessary adjunct functions include Provide Backup Power and Provide Load Distribution. These functions are all subordinate to the Provide Power function discussed in Section 3.3.

The normal way of providing electrical power is by a generator powered by mechanical torque from each engine. This engine produces alternating current (AC) power, which must be converted to direct current (DC) power for the DC electrical components. The DC power is also used for recharging batteries for backup power. Similarly, the DC power from the batteries is converted to AC power to provide power to the AC components which are needed during emergency operation: that is, they are part of the MMEL discussed in Section 5.4.

Load distribution is achieved by analyses during the synthesis phase. The objective of load distribution is to assure a balance of electrical loads so that no elements of the electrical system are overloaded when some components fail.

The electrical power subsystem synthesis is replete with constraints. Typical are the limitations on emitted EMI discussed in Section 5.10. EMI can be limited by sufficient cable shielding. Shielding also protects other components against surges in electrical power. All electrical power elements must be grounded, namely, to the airframe.

Other electrical power sources include the auxiliary power unit (APU) discussed in Section 9.7 and the ground power unit. Both of these may be used to provide electrical power when the aircraft is on the ground and to assist in engine starting.

A major top-level trade-off is the method of power generation. This trade-off is top-level because it trades the mechanical compensation for engine speed variation with electrical compensation. There are three major ways: First, the traditional method is called a constant-speed drive (CSD). With CSD the variation in engine speed is reduced to a constant generator speed entirely mechanically with a resulting mechanical complexity. In the second method, integrated drive generator (IDG), the mechanical speed reduction and generation functions are combined into a single unit. In the third method, variable-speed constant frequency (VSCF), the mechanical drive is allowed to vary while the generator produces a constant frequency output through electrical conversion. Factors affecting the final selection include efficiency, cost, reliability, and weight.

Shipside lighting (ATA 33-30, -40, -50)

Lighting functions are simple: Provide Exterior Lighting, Provide Lighting for Cargo and Bays, and Provide Exterior Emergency Lighting. All lighting functions can be said to derive mainly from the Communicate Data/Information function discussed in Section 3.3. Included are flight deck instrument panel lights, warning and caution lights, flood lighting, and other flight deck lighting. Requirements for flight deck lighting should consider the lighting level over the entire surface of the illuminated objects.

The lighting subsystem must provide lighting to cargo and service areas. As for the flight deck, the human factors aspects should be considered in determining the lighted areas and lighting intensities, that is, the actual lighting required for flight and service crews to perform their tasks.

Exterior lighting includes landing and taxi lights, anti-collision and position lights, ground floodlights, logo lights, wing and engine nacelle scan lights, and emergency lighting.

9.4   Interiors Segment

Most of the requirements associated with the interiors segment flow from the top-level function, Provide Passenger and Crew Accommodations discussed in Section 3.3. However, we will see below that this segment satisfies other emergency, information, and lighting functions as well.

Crew accommodations (ATA 25-10)

The key crew accommodations functions are to provide crew seating, storage, and equipment and furnishings. Other functions are to provide life support and evacuation.

The flight deck is a prime example of the integration and synthesis of several segments at a higher level than any individual segment. The flight deck should be laid out with many considerations in mind. Other segments, besides the interiors segment, with major assets in the flight deck include the avionics, mechanical, environmental, propulsion, airframe, and electrical segments. All of these elements should be integrated from a top-level point of view with a single purpose in mind: to allow the flight crew to fly the aircraft safely. The interiors segment plays a major part in this synthesis. The interiors segment is responsible for assuring the comfort of the pilots and the convenience of all flight deck equipment.

We saw in Section 7.4 that quality function deployment (QFD) provides an excellent methodology for transforming the qualitative requirements for interiors, such as passenger comfort, into an actual concept which meets customers' expectations.

Passenger accommodations (ATA 25-20)

The primary passenger accommodations functions are to provide passenger seating, provide passenger entertainment, provide storage for passenger items, and provide fixed interior items. In addition, the interiors segment provides acoustic and thermal protection with the sidewalls.

The most important considerations in the passenger cabin synthesis pertain to human factors and safety. We have seen in Section 5.5 that human factors determine the requirements for comfort, posture, reach, controls, and convenience. Safety considerations described in Section 10.2 drive many of the cabin requirements, for example, for aisle spacing, flame resistance, and many other factors. These factors are all clearly laid out in the FARs. As for the flight compartment, the interiors segment must provide life support, fire protection, and evacuation capabilities.

A key feature needed in passenger accommodation synthesis is *flexibility*. That is, the interior design should be flexible enough to satisfy the individual needs of various airline customers. Flexibility can be achieved through component modularity. Through modularity various components, for example, seats and overhead storage racks, can be designed to be rearranged in different customer configurations.

Among all segments, the interiors segment is very largely driven by direct customer requirements. Many interior items and materials are directly required by the airline customer. Of course, these direct requirements do not relieve the interiors engineer from a rigorous investigation of the interfaces with other elements as described in Chapter 6.

Water, waste, lavatories, galleys, and plumbing (ATA 25-30, -40, -38)

The principal functions of this subsystem are Provide Lavatory Capability and Provide Galley Capability. Subordinate functions include Provide Water and Provide Waste Disposal.

We have combined these five interiors subelements into a single subsystem because of the integral nature of their operation. Both the lavatories and the galleys utilize water and dispose of waste, and they are interconnected by the same plumbing system. Hence, in order to minimize the weight of the whole subsystem, it should be designed as a unit and not as individual pieces.

The modularity concept discussed in Section 9.4 is of particular importance to this subsystem because the movement of lavatories and galleys is difficult and involves many interfaces, primarily with the airframe and electrical subsystems and also many internal plumbing interfaces.

Emergency provisions (ATA 26-60)

The functions of this subsystem are to provide evacuation, flotation, interior fire extinguishing, and miscellaneous emergency capabilities. For evacuation, slides, life rafts, and assist lines for evacuation over the wing are provided. Miscellaneous emergency equipment includes fire axes and flash lights.

Signs and lights (ATA 33-10, -20)

This subsystem provides lights for passengers and crew and also provides the escape lights for evacuation. It also provides the placards for instructions and warnings. Considerations such as placards in appropriate languages, readability, and visibility are important.

Interior design

Interior design is one of the major factors in aircraft sales. Yet it is one of the most difficult to implement from an SE point of view. Such functions as Provide Passenger Comfort and Provide Passenger Entertainment are difficult to specify in verifiable terms. QFD described in Section 7.4 is a valuable tool in such circumstances. QFD allows the manufacturer, in cooperation with airline customers, to identify specific verifiable attributes of the interior design which will meet the needs of the customer. These attributes might include seat width and spacing, television viewing angles, and passenger reach requirements for overhead storage. QFD will also assist in providing a cockpit layout pleasing to pilots.

9.5   Mechanical Segment

Flight controls (ATA 27)

The major function of the flight control subsystem is Transfer Pilot and Autopilot Commands to Control Surface Deflection. An adjunct function is Provide Load

Feel, which is a subfunction of the Command and Control Aircraft function shown in Figure 3.6.

Typical subsystem-level trade-offs include: hydraulic vs. electrical (Fly-by-Wire) control. Another advanced option is Fly-by-Light (FBL) described in Section 2.4. With this option, control signals are transmitted by fiber-optic lines. Hydraulic power is generally more reliable and lighter than electrical. However, electrical systems can eliminate complex cable systems. Since the FARs require redundancy in control systems, both techniques can be used. In addition to these methods, propulsion controlled aircraft (PCA) present another control technique discussed in Section 9.6 for redundancy.

For control surface actuation subsystems, electrical, hydraulic, and pneumatic actuators are employed. Trade-offs to be considered include redundancy vs. dispatch reliability.

Hydraulic power (ATA 29)

The principal hydraulic function is Provide Hydraulic Power. Of course, hydraulic power is itself a solution. Therefore, the top-level trade-off is electrical vs. mechanical vs. hydraulic power. Auxiliary hydraulic power is a redundancy requirement for safety. Since hydraulic pumps are driven by mechanical torque supplied by the power plant, another top-level trade-off is autonomous vs. power plant driven hydraulic power. Some aircraft employ self-contained electrical motors and hydraulic pumps near the control surfaces they are moving. This is also subordinate to the Provide Power function discussed in Section 3.3.

A subsystem-level trade-off is hydraulic pressure vs. actuator size. Higher pressures can result in smaller actuators with a resultant greater risk.

Auxiliary hydraulic power may be provided by a hydraulic pump run by an electric motor powered by the APU. This same pump can be powered by the ground electrical power system.

Landing gears and brakes (ATA 32)

The landing gears and brakes functions emanate from the Provide Ground Movement function described in Section 3.3. In addition, this subsystem may receive requirements derived from the Generate Aero Forces function described in Section 3.3. Primary subfunctions are: Provide Braking, Provide Carriage, and Provide Steering. Other subfunctions include Provide Retraction and Extension, and Provide Access and Cover for Gear Retraction and Extension.

Gear sizing is driven by the requirement for aircraft rotation. Stroke length is driven by the required energy absorption on landing. Backup gear retraction can either employ the free-fall technique or the use of an auxiliary hydraulic actuator. As usual the primary figures of merit for all trade-offs are cost and weight.

Top-level trade-offs involving the landing gears and brakes are, first, main vs. nose gear steering and, secondly, the determination of the number of wheels.

The decision to employ main gear steering is determined by the required turn radius of the aircraft and by wheel loading. The number of wheels is determined primarily by the aircraft gross weight. The number of wheels is traded off against the tire size, which is determined by aircraft space limitations. These factors are also traded off against the scrub angle of the wheels (the angle between the plane of the wheel and the direction of motion). Large scrub angles cause increased tire and strut seal wear and limit the ability to tow the aircraft.

Another top-level trade-off is retraction time vs. aerodynamic performance. Quick retraction times require greater hydraulic power, while slower retraction times result in a reduced climb gradient. Another trade-off involving aerodynamic performance is the decision to use a wheel well cover or not. All major aircraft use wheel well covers to reduce drag. However, some smaller aircraft have open wheel wells.

The three most common landing gear arrangements are:

1. *Conventional tail wheel*   This configuration is not used on any modern jets. Its main advantage is that it is the lightest option. However, its disadvantages are that it has bad ground handling qualities and requires a fuselage inclination.
2. *Bicycle gear*   This configuration is the easiest for gear stowage. However, it is the hardest to land and maintain the proper wing angle of incidence.
3. *Tricycle gear*   This is the most common gear arrangement on modern subsonic jets. It has good handling characteristics, primarily because the main gear is near the center of gravity. It provides the largest moment arm for fuselage rotation.

9.6   Propulsion Segment

The primary top-level function supported by the Propulsion Segment is the Provide Total Impulse function in Figure 3.6.

Fuel subsystem (ATA 28)

The two primary functions of the fuel subsystem are Provide Fuel to the Engines and Provide Fuel to the APU. Functions which follow from these are Provide Fuel Storage, Provide Fire Protection, Provide Fueling and Defueling, Provide Dumping, and Provide Fuel Control. We saw in Section 9.1 that the fuel system has also inherited another function, namely, Provide Wing Anti-Icing. Other aspects of the power plant subsystem described in Section 9.6 also put demands on the fuel consumption, namely, electrical and hydraulic power and bleed air. Like the wing, the fuel subsystem also provides wing load alleviation because of its weight. The fuel system can also be used to cool the engine oil.

Another aspect of the fuel subsystem is fuel management which emanates from the Manage Fuel function discussed in Section 3.3. The primary function of fuel management is to maintain fuel circulation to prevent the formation of ice in the fuel. Other key functions include the control of the center of gravity (c.g.) and the cross-feeding of fuel between the engines to maintain a balance in the availability of fuel and to prevent a lateral imbalance.

The fuel system synthesis is inextricably linked to the top-level aircraft synthesis. We have seen in Section 8.2, Item 7 that the fuel weight is one of the earliest top-level *derived* requirements. This requirement immediately flows to the fuel subsystem for sizing that subsystem. Fuel storage is normally limited to the wing because of the inherent efficiency of integrating the wing structure and the fuel storage. Hence, major trade-offs are required between fuel storage requirements and aerodynamic performance. Longer aircraft ranges demand more fuel. This demand is compatible with thicker wings and/or wings with increased chord length. If the wing span is limited by airport constraints, then the aspect ratio discussed in Section 8.2 needs to be traded against the larger chord lengths. These are the trade-offs which result from fuel requirements.

Pylon (ATA 54 through 54-80)

The pylon has two primary functions: Provide Support for Engines and Provide Conduit for Subsystem Functions to and From Engine. The pylon can also be thought of as part of the airframe segment; however, because of its importance to the engine, we have included it as part of the propulsion segment. The pylon is an important element whether the engines are wing-mounted or body-mounted, a top-level trade-off. The pylon is also an important element in the Provide Fire Protection function. Another function of the pylon is Provide Aero-Elastic Damping to prevent aerodynamic flutter.

Power plant (ATA 71)

The primary power plant function is Provide Thrust discussed in Section 3.3. However, a number of other functions have been imposed on the power plant, namely, Provide Bleed Air to Pneumatic Subsystem, Provide Mechanical Torque for Hydraulic Power, and Provide Mechanical Torque for Electrical Power. Provide Fire Protection is a necessary adjunct function. All of these functions are subordinate to the Provide Total Impulse function discussed in Section 3.3.

Other functions which have emerged through practice are Alleviate Wing Load and Provide Lift Augmentation. The propulsion system alleviates wing loads by off-setting the lift load with its weight. It provides lift augmentation through strakes which cause the turbulent boundary layer to stay attached and thus improves the maximum lift coefficient, C_{Lmax}, during take-off. Another secondary, but important, function is Provide Noise Attenuation. The power plant achieves this function by integrated noise treatment involving many components. Berry (1993) cites the inlet, fan, compressor, fan nozzle, burner, turbine, and primary nozzle as being the principal noise sources.

An important subfunction is Start Engine. To accomplish this function, the engine needs a source of compressed air. On the ground this air can be provided either by the ground equipment or by the APU. In flight the engine can be restarted by free air or by the cross-feed of air from another engine.

Of the four primary functions, the most important is Provide Thrust. All of the others have been arrived at by trade-offs over decades of experience. Although pneumatic air, hydraulic power, and electrical power could be obtained by alternative means, experience has shown that the power plant is the most efficient source. With good SE, these decisions will be revisited as technology and economic factors shift.

We have seen in Section 8.2 that the number of engines is a top-level trade-off. For every mission (payload, range, speed) there will be an optimum number of engines which can be found through the trade-offs described in Chapter 8. In addition, FAR constraints will establish the minimum number of engines, thrust, and predicted reliability required for emergency conditions, such as engine-out over water.

Another top-level trade-off is engine placement. Wing-mounted engines result in a lighter wing weight. In addition, their location farther away from the fuselage makes it easier to maintain lower interior noise levels. However, the advantage of rear-mounted engines is that the aircraft is easier to control in an engine-out situation.

For wing-mounted engines, a key trade-off is the placement of the engines on the wing. Although engines nearer the tip are preferred for wing loading, engines near the fuselage are superior for controllability in a single engine-out condition.

Berry (1993) provides a comprehensive set of design considerations and trade-offs for the propulsion segment.

Thrust management (ATA 76)

The Manage Engine Thrust function is the process of applying forward or reverse forces as required for each phase of flight and landing roll-out. The key subsystem-level trade-off is manual vs. automatic thrust management. In automatic thrust management the thrust is managed by a computer. The pilot's role is to set the throttle at various appropriate aircraft phases, such as take-off, climb, cruise, descent, or landing. A secondary possible thrust management function is Provide Aircraft Control. By automatically managing the thrust of the engines, the thrust management subsystem can control the aircraft in roll, pitch, and yaw.

Propulsion monitoring

The propulsion monitoring subsystem maintains a constant vigilance of the propulsion segment's health for maintenance. In addition, it assures that all elements of the segment adhere to their red-line limits for safety.

9.7 Auxiliary Segment (ATA 49)

An APU is not an absolute necessity on an aircraft. However, most commercial aircraft have an APU to provide functions which may not be available at some airports. In addition, it provides a redundant source of power for aircraft functions in the event the engines are inoperable. Its primary functions are Provide Auxiliary Electrical Power, Provide Pneumatic Power for Engine Starting discussed in Section 9.1, and Provide Auxiliary Pneumatic Power for Air Conditioning. The electrical power function is provided as a backup to the primary electrical power source in flight. The pneumatic functions are primarily ground functions. The auxiliary power functions are subordinate to the Provide Power function discussed in Section 3.3.

As for the power plant, the Provide Fire Protection for the APU is a necessary adjunct function. The primary reason for the APU's existence is to provide the autonomous starting capability at airports where no external ground power sources are available. This capability emanates from a very top-level aircraft requirement for autonomous starting capability, if required for the route structure of the aircraft. The other functions, electrical and pneumatic power, are backups to the primary subsystems. These functions emanate from safety and operational considerations. Other auxiliary functions could be performed by the APU, such as Provide Hydraulic Power, if necessary.

9.8 Airframe Segment

All of the airframe components listed below have many aspects in common. They are subject to similar performance requirements and constraints, and many of the same trade-offs are the same. Functions in common include Sustain Loads and Maintain Aerodynamic Profile. These functions support the top-level functions, Provide Aerodynamic Performance discussed in and Maintain Structural Integrity both discussed in Section 3.3.

A key trade-off for the airframe is between conventional materials, e.g. aluminum, vs. newer composite materials. Composite materials provide the potential of lighter weight and greater strength. These benefits should be traded against higher recurring costs and possibly higher maintenance costs. All of these factors fit into the DOC discussed in Section 8.6. That is, if the reduced weight and fuel used with composite materials more than compensate for the increased manufacturing and maintenance costs, the DOC will decrease, thus making the aircraft more economical to operate and therefore more marketable.

Constraints discussed in Chapter 5 are particularly important in airframe design. Environmental constraints pertaining to corrosion and shock loads discussed in Section 5.6, for example, apply to all airframe elements. Especially important are durability requirements. Each airframe member should be shown to last a given number of years without failure. Within the category of qualitative safety requirements discussed in Section 10.2, the principle of fail-safe design applies. That is, if one structural member fails, due to cracking, for example, another should not exceed its limit load, that is, two-thirds of its ultimate load.

Fuselage (ATA 57)

In addition to the two primary functions discussed above, the fuselage must perform the following functions: Maintain Pressure, Provide Access for Ingress and Egress; Provide Cargo Loading; Provide Space for Passengers, Crew, Cargo, and Subsystems; Provide Cargo Fire Protection; and Provide Pilot Visibility. Provide Pilot Visibility is a subfunction of the Communicate Data/Information top-level function discussed in Section 3.3.

Of these, the Maintain Pressure function is perhaps the most demanding. For the HSCT, this function will be one of the most critical from a safety point of view since loss of pressure will most likely be a catastrophic event, unlike with subsonic aircraft.

Empennage (ATA 55)

In addition to the primary functions, the key function of the empennage, also called the tail, is Provide Aerodynamic Control. This function includes both the aerodynamic stability which the vertical and horizontal stabilizers provide to the whole aircraft as well as the vertical and lateral control provided by the vertical (rudder) and horizontal (elevator) moving surfaces. Of course, with the implementation of propulsion control discussed in Section 2.4 there would be no need for the moving surfaces. This trade-off is anticipated to result in a safer aircraft because of the reduced number of control systems.

A key tail top-level trade-off is the horizontal tail placement. The location of the horizontal tail is dependent on the location of the engines. The T-tail configuration is almost certainly required for fuselage-mounted engines, while fuselage-mounted horizontal tails are preferred for wing-mounted aircraft.

Wing (ATA 57)

The primary function of the wing, it is assumed, is to provide the lift that keeps the aircraft aloft. However, this function is just part of the Maintain Aerodynamic Profile discussed above. The wing also serves to maintain a minimum aerodynamic drag. The use of winglets has increased in recent years to accomplish both of these functions.

In addition to the two principal functions above, the wing must perform the following functions: Transfer Loads to Fuselage, Provide Storage for Fuel, Provide Aerodynamic Control, and Provide Space for Subsystems. We saw in Section 8.2 how key wing requirements (wing area, aspect ratio, sweepback angle, and taper ratio) are the subject of aircraft-level trade-offs. The primary wing design trade-offs are driven by the Provide Storage for Fuel function as discussed in Section 9.6. Ailerons, strakes, slats, and spoilers have been used for many years to facilitate the Provide Aerodynamic Control function.

9.9 Allocation to Software

From an SE point of view, software is exactly the same as any other part of the aircraft. It has functions, and it has requirements. Every subsystem listed above will have software. So, we have not listed software here to imply that it is a separate subsystem. We only want to show that it should be included it in the whole SE scheme during the formulation phase of the aircraft or any part of it. Section 10.3 presents a more extensive discussion of the development and certification requirements and constraints that are normally imposed on software.

9.10 Subsystem Constraints

For subsystem constraints, the general principles of Chapter 5 apply. However, it is worth emphasizing the following points: First, the general industry and regulatory design constraints which were laid on the aircraft as a whole apply to the individual subsystems as well. Hence, their inclusion in the design is just as subject to scrutiny in design reviews and verifications. Secondly, the allocation procedures (4.7) apply to all subsystems. The most notable ones are weight and dispatch reliability, among others.

10

Certification, Safety, and Software

The objective of the certification process is to substantiate that the aircraft and its systems comply with applicable requirements.

ARP 4754A (2010, p. 76)

The certification process, as indicated by the quotation above, is focused on the safety aspects of the aircraft development as determined by such agencies as the Federal Aviation Administration (FAA) and the Joint Aviation Authorities (JAA). These requirements are spelled out in documents, such as Federal Aviation Regulations (FARs) and Joint Airworthiness Regulations (JARs). The FAA has taken a major step in incorporating SE principles into the certification process by the publication of ARP 4754 (1996) and later in ARP 4754A (2010). This document is a set of guidelines compiled and published by the Society of Automotive Engineers (SAE) in cooperation with the FAA. Hence, it represents a look into the future of certification rather than present day practices. However, this look is extremely important from an SE point of view since it demonstrates the FAA's and SAE's commitment to the SE process.

A second aspect of certification which supports SE principles is verification. While the certification process focuses on those verification techniques which support airworthiness, it nonetheless forms part of the total SE verification process dedicated to the goal of 100 percent verification.

This chapter does not intend to present a complete description of the certification requirements and process, but rather to outline basic features of the process and to show how this process is integrated with and is mutually compatible with the SE process. Thus, the reader should not rely on the contents of this chapter as a source of certification requirements but rather should obtain original source material or contact the appropriate regulatory agencies. As stated in the Preface, this book does not intend to replace the official standards and guidelines, or to be a definitive interpretation of them, but rather to be a *pointer* to them and to show how they can be seen in the SE context, and how these processes can be adapted to the commercial aircraft domain.

Although safety is the primary focus of certification, the certification requirements affect the design, design processes, and management processes beyond the realm of safety. For example, aircraft manufacturers must submit a description of the features of their design processes and configuration management processes, which are used in both safety related and non-safety related requirements, design, and verification processes.

The terminology used in ARP 4754A sometimes differs slightly from the terminology used in SE. ARP 4754A uses the term *system* in the traditional aircraft sense, that is, to mean a subsystem in the SE context. We have used SE terminology here, that is, *subsystem* where appropriate. ARP 4754A refers to passenger safety as a *function*. Many systems engineers, the author included, would refer to safety as a constraint or specialty requirement. Since SE has not totally standardized its own terminology, these differences cannot be regarded as significant. More important than the terminology is the assurance that the SE process is complete and that all of these considerations are included and analyzed properly.

This chapter also discusses the development and certification of systems with software. The reason for the grouping of software with certification and safety is that software development is a critical safety element and contains its own certification process requirements within the overall aircraft certification process.

Finally, it should not be inferred that the discussion of the certification processes in this chapter forms a definitive interpretation of the regulatory documents discussed in it. The reader is advised to refer to those documents when engaged in actual certification processes. The purpose of this chapter is only to show that the certification processes are compatible with SE principles.

10.1 Certification

This section will discuss how the certification requirements outlined in ARP 4754A are compatible with the SE process. ARP 4754A was written primarily, but not exclusively, with electronic subsystems in mind; however, the principles apply to any type of system or subsystem. Table 10.1 shows how the required elements of certification data are in agreement with SE principles.

10.2 Safety

Safety is the primary concern of the certification process. The term *safety* covers both safety related design constraints as well as the quantitative safety requirements. For certification, safety constraints consider both the availability (continuity) as well as the integrity (correctness of behavior) of the function. In other words, safety analysis is concerned both with *whether* a function is performed as well as with the risks inherent in not achieving the desired performance of the function.

Quantitative safety requirements

All quantitative safety requirements result from the FAR 25.1309 requirement, which states:

> The airplane equipment and associated components, considered separately and in relation to other [subsystems] must be designed so that (1) any catastrophic

failure condition is (a) extremely improbable, and (b) does not result from a single failure; and (2) any hazardous failure condition is extremely remote, and (3) any major failure condition is remote.

Table 10.1   Compatibility of the SE and certification processes

SE Elements	Certification Aspects
Functional analysis (Chapter 3)	Aircraft-level functional requirements, allocation of aircraft functions to systems, functional hazard analysis (FHA) discussed in Section 10.2
Requirements development and allocation (Chapter 4)	Requirements categories discussed in Section 4.9, allocation of item requirements to hardware and software, preliminary safety assessment (PSSA) discussed in Section 10.2, validation plan and data
Synthesis (Chapters 7–9)	Development of system architecture, common cause analysis (CCA) discussed in Section 10.2, system implementation
Verification and validation (Chapter 11)	Verification data, system safety analysis (SSA) discussed in Section 10.2, inspection and review process
SE management (Chapter 12)	Certification plan, development plan, configuration management plan and data, process assurance plan and evidence, certification summary

This requirement can be paraphrased to say that the hazard probability must be inversely proportional to the hazard severity. This FAR sets the total maximum allowable probability (MAP) of a catastrophic event at 10^{-7}. This requirement means that for each of 100 possible events, for example, the probability cannot exceed 10^{-9}.

Functional Hazard Assessment (FHA)

The functional hazard assessment (formerly called functional hazard analysis) is an integral part of the safety process. The FHA is tied to the SE concept of functional analysis discussed in Chapter 3. It is a systematic and comprehensive examination of a subsystem's functions. Its purpose is to determine potential hazards a subsystem can cause or contribute to, not only if it malfunctions, but also in its normal operation. The FHA provides the results of this examination as an assessment at the overall system level. The FHA requires a hazard assessment for each aircraft function and for each *combination* of aircraft functions. Hence, the FHA has introduced the philosophy that hazards may be caused by combinations of functions.

The development of the functions themselves may seem like a formidable task, and it is. The saving grace is that most of the task only has to be done once since most of the higher-level functions are the same from aircraft to aircraft. New functions will only have to be developed as new features are introduced onto aircraft. These functions will normally appear at lower levels of the functional hierarchy. This factor will tend to simplify the FHA.

Preliminary System Safety Assessment (PSSA)

The PSSA evaluates the proposed architecture and compares it to the failure conditions in the FHA described in Section 10.2 in order to determine the safety requirements of the subsystems or items. For example, the PSSA could determine the degree of redundancy required to mitigate a hazardous condition.

System Safety Assessment (SSA)

An SSA is a systematic, comprehensive analysis of the system functions to show that the safety requirements have been implemented in the design. The SSA evaluates all aspects of the system concept from a safety point of view. It reviews the functions, interfaces, event probabilities, failure conditions, combinations of functions, maintenance factors, and the results of all other analyses. Hence, the SSA is an integral part of the SE verification process as verification by analysis described in Section 11.2.

Common Cause Analysis (CCA)

The purpose of the CCA is to ensure the independence of subsystems which have faults in common. The potential for common faults is most common in subsystems which rely on redundancy or on the same software which is used by more than one subsystem. Hence, CCA falls within the synthesis phase of development described in Chapter 7.

The CCA performs a zonal safety analysis to ensure multiple subsystems in the same aircraft zone do not interfere with each other. The CCA performs a particular risk assessment to examine events and influences which may affect more than one subsystem. The CCA performs a common mode analysis to determine whether or not subsystems are truly independent.

Safety and the SE processes

Table 10.1 shows that the safety assessment required by the certification process is compatible with SE process. FHAs described in Section 10.2 are conducted at both the aircraft and subsystem levels. The PSSA described in Section 10.2 inserts safety squarely in the middle of the requirements analysis early in the SE process. The SSA described in Section 10.2 contributes to the verification of the implemented system.

Organizational safety

The concept of organizational safety is a product of SE in its broadest sense because it considers not only the operational aspects of the aircraft but also the development as well. Studies have shown that the common factor in major catastrophes, such as the Space Shuttle *Challenger* and the North Sea disaster, was not technological but rather organizational as described by Paté-Cornell (1990). Organizational factors identified were: time pressures, failure to observe warnings of deterioration and signals of malfunction, lack of an incentive system to handle properly the trade-offs between productivity and safety, failure to learn from mistakes and motivate reporting of problems, and lack of communication and processing of uncertainties. It is the responsibility of the program management in cooperation with the Chief Systems Engineer (CSE) to assure that these factors are minimized as discussed in Section 13.2. Although organizational safety is not specifically mentioned in either standard SE or certification guidelines, it could be considered to be an implied aspect of process assurance shown in Table 10.1. In addition, the evidence discussed above shows that it warrants increased attention.

Qualitative safety

Qualitative safety requirements are all those requirements necessary to meet the quantitative safety requirements of Section 10.2. Many of these requirements result from the PSSA described in Section 10.2, the CCA also described in Section 10.2, or the FHA also in Section 10.2. In each case the requirement must be verifiable as shown in Section 4.1.

   The purposes of the qualitative safety requirements are typically: to restrict the severity of failures, to assure that one failure does not cause another failure, or to assure redundancy of critical functions. A typical example is the requirement for double retention of bolts. That is, if a bolt fails, the structural load will be sustained by another bolt.

10.3   Software Development and Certification

Software constitutes a major portion of the development costs of an aircraft. In addition, it is a major focus area in certification. Hence, software development should be approached as methodically as the rest of the aircraft; that is, the software should be considered part of the total aircraft, not a separate system to be developed separately. The basic principles discussed in this book (for example, performance requirements, constraints, synthesis, and verification) apply equally to software and firmware as to hardware.

Software has two characteristics which make it unique and require special treatment: The first is its importance with respect to safety. Virtually every subsystem on the aircraft contains software, avionics more than the others. The second is the difficulty in developing and verifying it. These two aspects make it a critical factor in safety studies and in certification.

Specific methodologies and standards have been produced for the development and certification of systems with software. Notable among these is RCTA/DO-178B (1994), which is the primary standard for software for commercial aviation. It is not the purpose of this section to describe the software development and certification process, but rather to show how that process already employs SE principles and therefore is compatible with the SE process. Following are a few of the principal considerations.

Software relation to system

The development of the system and the software is a two-way street. The following aspects of the system must be developed and provided to the software: First, the system requirements are allocated to the software in accordance with the SE allocation process described in Section 4.5. Secondly, the established software levels are assigned described in Section 10.3, Item 1) in accordance with the criticality and risk of the particular software application. Other software design constraints described in Section 10.3 are assigned, and hardware is defined. Similarly, the software requirements and architecture are introduced as an integral part of the entire system. Error sources have either been identified or eliminated for safety assessment, and fault containment boundaries have been established.

The software life-cycle is analogous to the aircraft and SE life-cycle functions described in Section 3.1. First, there is the planning process which lays out the entire software development process. Secondly, there is the development process itself. Finally, there is a set of activities which should be conducted concurrently. These include verification, configuration management, quality assurance, and the certification liaison process. All of these processes occur for the aircraft as a whole. Table 10.2 shows how the elements of software development and certification fit into the SE process.

Most importantly from an SE point of view, the certification authorities do not consider software to be a separate entity from the rest of the aircraft. Therefore, the aircraft manufacturer is obliged to submit certification compliance evidence in a total aircraft context. Hence, the certification basis as shown in Table 10.1 for the entire aircraft must include the software aspects of certification. The certification authorities will consider the plan for software aspects of certification along with other aircraft data required for certification. The software accomplishment survey will be an essential ingredient in certification compliance.

Software constraints

Like other subsystems in the aircraft architecture, software must adhere to prescribed constraints. The following are the principal software constraint categories. These constraints are normally documented in the system, that is, the aircraft specification because they affect the quality of the entire aircraft. These constraints also apply to firmware:

1. *Software level* In SE terms, all software requirements are *derived* requirements. That is, functional requirements are allocated to both hardware and software after a design concept has been selected and it has been decided where the software will fit into the system. When it has been decided where the software is going to fit in, what functions the software will perform, and what the hazard categories of the functions are from the system safety analysis (SSA) described in Section 10.2, a software level can be assigned to the software item. These levels are labeled A, B, C, D, and E and establish the level of quality and testing necessary for the software. Software levels must be approved by the certification authorities shown Table 10.2.

Table 10.2   Compatibility of the SE and software development and certification processes

SE Elements	Software Development and Certification Aspects
Functional analysis (Chapter 3)	Software functional analysis, system functions allocated to software, criticality of software functions through FHA discussed in Section 10.2.
Requirements development (Chapter 4)	Software requirements development, allocation of system requirements to software, software levels discussed in Section 10.3, Item 1), design constraints, hardware definition to software, software requirements, software redundancy requirements through PSSA discussed in Section 10.2.
Synthesis (Chapters 7–9)	Software synthesis, software architecture, error sources identified and eliminated, fault containment boundaries established, software independence through CCA discussed in Section 10.2.
Verification and validation (Chapter 11)	Software testing and analysis, evaluation of software implementation SSA discussed in Section 10.2.
SE management (Chapter 12)	Software planning, software configuration management (SCM), software configuration index (SCI), software quality assurance (SQA), plan for the software aspects of certification (PSAC), liaison with certification authorities.

2. *Protection requirements*   It is necessary to protect software functions from each other and from other functions. Software protection is an essential requirement in software development shown in Table 10.2 and usually takes the form of partitioning by time and space.

3. *Software standards*   Software standards determine the requirements for programming language, control structures, and other design constraints. These standards are also another part of software development shown in Table 10.2.

4. *Memory and timing consumption*   The memory requirement specifies the required excess memory for any given application.

5. *Memory protection*   Memory protection requirements are developed to protect the computer memory from influences, such as electrical power source transients or EMI, HIRF, or lightning generated transients.

6. *Program storage and integrity*   It is the purpose of program storage and integrity requirements to specify the reliability, availability, and other features of data integrity when it is stored in memory.

7. *Computer anomalies*   The computer anomaly requirements specify the timing and detection of potentially unsafe conditions in a computer and the change to a safe condition. Littlewood and Stringini (1992) emphasize the difficulty in predicting these anomalies and argue that the use of computers for performing complex decisions should be avoided.

8. *Fail-safe design*   The requirements for a fail-safe design specify the principles to be employed to ensure a safe design. The objective of this requirement is to make the design less sensitive to failure modes.

9. *Robustness*   The robustness requirements specify the ability of the software or firmware to operate in the face of external failures or invalid inputs.

10.4   Commercial Aviation Safety Team (CAST)

According to the overview, the Commercial Aviation Safety Team (CAST 2011) is an international partnership of aircraft manufacturers, employee groups, regulatory authorities, and aircraft operators working together to enhance safety in the commercial aviation domain. CAST makes recommendations to all elements of the aviation domain for ways to enhance safety. Their record is impressive. According to the FAA (2013), just one of the CAST members, CAST recommendations have resulted in a decrease in the fatality rate of 83 percent in the United States.

Recommended safety enhancements for manufacturers

CAST periodically publishes safety enhancement summaries detailing recommended enhancements to aircraft intended to improve safety. Enhancements include actions by manufacturers, regulators, and air carrier associations. Following is a summary of such a safety enhancement sheet published by CAST (2012) and (2014); the summaries below are highly condensed and paraphrased; refer to the original for exact wording and for responsibilities. Enhancements underway at the time of publication of the fact sheet are not listed here. In addition, the following list contains items that are in progress and not completed at this time.

Category: Controlled Flight into Terrain (CFIT)

Enhancement: Terrain Avoidance Warning System (TAWS)

- Install TAWS on all newly manufactured aircraft.
- Retrofit TAWS on existing aircraft.
- Institute system to support TAWS including installation, maintenance, and training.
- Develop standard operating procedures (SOP) for flight deck crew members.
- Include vertical angles in instrument approach procedures.
- Incorporate a digital elevation model to determine minimum vectoring altitudes (MVA) to reduce TAWS alerts.
- Develop procedures to provide better separation from terrain at selected sites.

Enhancement: Runway lighting

- Install visual glide slope indicators (VGSI) on all runways used by carriers.

Enhancement: Distance measuring equipment (DME)

- Install DME on older aircraft.

Enhancement: Improved area navigation (RNAV) procedures

- Include vertical guidance in RNAV procedures.
- Incorporate required navigation procedures (RNP) technology to allow for more precision landings.
- Incorporate advanced precision approach procedures into the Next Generation Air Traffic System design.
- Institute plan for periodic check of the minimum safe altitude warning (MSAW) system.

Enhancement: Proactive safety plans—Flight Operational Quality Assurance (FOQA) and Aviation Safety Action Plan (ASAP)

- Develop voluntary procedures and protocols for trends and corrective actions.

Enhancement: Improved crew resource management (CRM) procedures

- Promote SOP for CRM.
-

Enhancement: TAWS Improved Functionality.

- Install GPS sensors on all aircraft.
- To achieve the capability to update terrain databases for operators.
- Update TAWS underlying algorithms.
- Evaluate TAWS features currently not used.

Enhancement: Improved CFIT training

- Add CFIT training to all required air carrier curricula.
- Add CFIT training for all controllers.

Category: Approach and Landing Accident Reduction (ALAR)

Enhancement: Airplane flight manuals (AFMs)

- Provide inspectors with latest AFM database.

Enhancement: Flight deck equipment upgrade

- Develop advisory material for checklists and alerts for new type designs.
- Incorporate FAA Human Performance Considerations in checklists.
- Provide automatic aural altitude call outs.

Enhancement: Implementation plan for aircraft design

- Ensure continuing airworthiness processes and incorporate risk management techniques; monitor fleet performance and prioritize safety critical threats and corrective actions. (See Chapter 15 for an elaboration of risk management considerations and Section 4.3 of the *FAA Systems Engineering Manual* (2014).)

Enhancement: Promote safety culture

- Promote safety culture for all chief executive officers (CEOs) and directors of safety (DOSs).
- Incorporate safety culture information in manuals.

Enhancement: Maintenance rules

- Re-emphasize maintenance rules for landing struts in maintenance programs.

- Re-emphasize maintenance rules for landing struts for maintenance suppliers.
- Increase oversight of maintenance procedures for suppliers.
- Include minimum equipment list (MEL) in maintenance procedures.
- Directs DOS to determine maintenance deficiencies.

Enhancement: Installation and improvement of flight deck equipment

- Implement electronic checklist and smart-alerting systems.

Enhancement: Flight crew training

- Implement flight crew training volutarily over ALAR topics.

Enhancement: Aircraft design

- Incorporate fault tolerant design principles.

This recommendation is in agreement with the resilience principles described in Chapter 16.

Category: Loss of control

Enhancement: Inform personnel and flight crew of policies and procedures

- Review processes for distributing essential information.
- Distribute information to flight crews and maintenance personnel.
- Include essential information in training programs and flight manuals.

Enhancement: Standard operating procedures (SOP)

- Publish and enforce and provide training for SOP for all phases of flight.

Enhancement: Risk assessment and management

- Implement methods to prioritize safety-related decisions.

This recommendation is in agreement with the methodology of Chapter 15 on Risk Management. However, this recommendation focuses on safety-related risks, while Chapter 15 discusses a broader range of risks.

Enhancement: Human factors and automation

- Compile a list of policies and procedures dealing with mode awareness and energy state awareness.

- Disseminate improved automation policies and procedures to operators and manufacturers. (The Billings (1997) list of rules for the interaction of humans and automation is a potential source of policies and procedures for this recommendation.)

Enhancement: New aircraft designs

- In the design of new airplanes incorporate angle of attack and low-speed protection, thrust-asymmetry compensation, and bank-angle protection.
- Airbus (2013) describes this technology called flight envelope protection and its implementation. This technology is among the new technologies discussed in Section 2.4.
- In the design of new airplanes incorporate features that will minimize thrust asymmetries, yield to manual force control when necessary, incorporate annunciations when necessary, and include low-speed protection.
- For all aircraft incorporate vertical situation displays in aircraft designs.
- For new aircraft incorporate displays and annunciations that will reduce accidents due to loss of control.
- For new aircraft designs which do not incorporate evaporative anti-icing systems, incorporate advanced anti-icing features.

Enhancement: Flight crew proficiency

- Develop voluntary program to improve flight crew proficiency.

Enhancement: Advanced maneuvers training

- Develop training program to prevent and recover from flight conditions outside the normal operating envelope.

Enhancement: Runway incursion

- Provide enhanced air traffic control (ATC) training and flight crew training for runway incursion (RI).
- Provide enhanced CRM training for air traffic controllers.
- Establish, document, and train SOP for commercial aviation ground operators, general aviation ground operators, tow tug operators, vehicle operators.
- Incorporate new technologies and new procedures that will improve situational awareness for air traffic controllers.
- Clarify procedures for air traffic control instructions and mandatory readbacks.

Enhancement: Turbulence

- Standardize methodologies for improved situational awareness and procedures during flight turbulence.

Enhancement: Uncontained engine failures (UEF)

- Implement advanced methods and technologies to detect potential engine defects. This item is in agreement with the concept of *latent* flaws discussed in Chapter 16.

Category: Maintenance

Other sources, for example Reason (1997, pp. 85–105), have concluded that faulty maintenance is a root cause of many aircraft and other accidents. The enhancements listed below address some of the deficiencies inherent in the maintenance process. Reason mentions American Flight 191 and Japan Airlines Flight JL 123 as examples.

Enhancement: Advanced circuit design

- Implement technologies to determine fault sources for maintenance.
- Implement advanced circuit breaker technology.
- Implement processes to certify advanced circuit breaker protection.
- Implement and certify advanced circuit breakers on in-service airplanes.

Enhancement: Maintenance policies and procedures

- Assure that work cards, procedures, and manuals are complete, accurate, available, and appropriately used.
- Assures that maintenance data are collected and reported to the original equipment manufacturer (OEM). This item is in progress.
- Identify and correct gaps in the maintenance process that may affect aircraft safety. This item is also in progress.
- Provide visible tagging of pitot tubes during maintenance when covered to enhance preflight walk-arounds.

Enhancement: Wrong runway departure

- Review existing signs and marking plans at high-threat airports, identify potential hazards, develop mitigation plans, and incorporate necessary changes.

- Incorporate wrong-runway operations in pilot and controller training programs and in Airport Layout Reviews, in policies for early take-off clearances, and in the installation of moving map displays and runway awareness systems.

Category: Cargo

Enhancement: Fire containment

- Develop standard fire suppression and/or containment systems.

Enhancement: Load training and SOP

- Plan and enforce contractor training for cargo loading.

Enhancement: Hazardous materials

- Implement improved regulations, methods, technology, and training for detecting and preventing hazardous materials.
- Develop systems to contain and suppress fires resulting from hazardous materials.

Enhancement: Processes and oversight of cargo-related issues

- Enhance legal processes for dealing with fires resulting from hazardous materials.
- Enhance oversight of cargo-related issues.
- Develop safety culture for dealing with cargo issues.

Category: Icing

Enhancement: Avionics

- Design and install smart pitch guidance systems.

Enhancement: Training

- Develop training material for dealing with engine events related to icing.

Category: Midair collisions

Enhancement: Terrain Collision Avoidance System (TCAS) policies and procedures

- Design B/C/D airspaces to make visual flight rules (VFR) more easily usable. (Note: B/C/D refers to defined airspaces in which instrument flight rules (IFR), special VFR, and VFR apply.)

- Conduct study to determine if TCAS alerts can be reduced or eliminated at high elevation airports. This item is in progress.
- Conduct study to determine if an adjustment of lateral or vertical air traffic separation minima would reduce or eliminate the number of TCAS alerts between IFR and VFR traffic. This item is in progress.
- Conduct study to determine whether current TCAS can support NextGen traffic levels. Create new strategy for NextGen. This item is in progress.

Category: Airplane State Awareness (ASA)

Enhancement: Low speed alerting

- Implement manufacturer service bulletins to provide low speed alerting on existing transport type designs.

Enhancement: Crew state awareness

- Specify non-standard, non-revenue flights for functional checks.
- Conduct risk assessment and develop guidelines for non-standard, non-revenue flights.
- Develop SOP to reduce flight crew loss of state awareness and develop training programs.
- Verify and validate crew capability with third party training providers in executing SOP for state awareness after receiving training.

This third party verification and validation is in agreement with the independent review principle described in Chapter 15, Risk Management.

- Implement standard practices with respect to upset (loss of control) or stall resulting from lack of state awareness; incorporate in simulators and in realistic scenarios.
- Emphasize control and stabilized flight in non-standard situations.
- Incorporate realistic training in go-around scenarios.
- Incorporate CRM in training.
- Incorporate virtual meteorological conditions (VMC) in displays.
- Incorporate bank angle alerting and recovery guidance displays on new aircraft and FBW programs.
- Study incorporation of bank angle protection and energy state cues in FBW, existing non-FBW systems, and out of production aircraft. This item is in progress.
- Study new technologies to improve state awareness. This item is in progress.
- Conduct research to improve state awareness; display in an intuitive manner; prioritize and escalate alerts; improve technology readiness levels (TRL) as shown in Table 7.1; this item is in progress.

- Conduct research to determine the benefits of using various levels of prototype advanced aerodynamic modeling of full stall characteristics to perform full stall recovery training; this item is in progress.
- Conduct research to assess crew performance with respect to state awareness; this item also includes attention-related performance; this item is in progress.

Conclusions

The following conclusions can be drawn from the above CAST recommendations: First, recommendations are just recommendations; they are not mandatory requirements. They will not be mandatory until the regulatory agencies make them mandatory. In some cases the manufacturers will implement them even if they are not mandatory. Secondly, most of the recommendations are procedural or administrative. Therefore, the cost of implementing them will be minimal. Hence, the benefit to cost ratio can be expected to be large.

10.5   Fatality Rate History

Figure 10.1 shows the fatal accident rate for Part 121 that is scheduled carrier, operations over more than 30 years. It is apparent that safety recommendations, such as those made by CAST, have had a significant impact on safety.

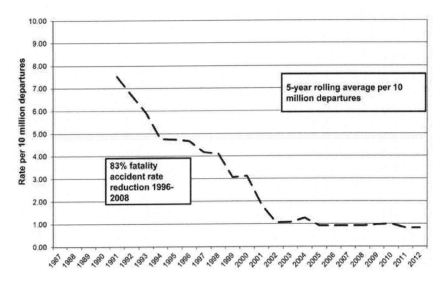

Figure 10.1   Fatal accident rate for Part 121 operations (scheduled carriers)

11

Verification and Validation

[Verification is] the evaluation of an implementation of requirements to determine that they have been met.

ARP 4754A (2010, p. 13)

[Requirements validation is] the determination that the requirements for a product are correct and complete.

ARP 4754A (2010, p. 13)

[Product validation is] actions to confirm that the behavior of a developed system meets user needs.

Stevens et al. (1998, p. 364)

Verification of requirements in the commercial aircraft industry is an extremely complex process. Verification of certification-related requirements is controlled by Federal Aviation Regulations (FARs) and thoroughly conducted and monitored as shown in Table 10.1. These requirements and the balance of the requirements are verified by a variety of methods, including ground tests, simulations, analyses, examinations, and flight test. Similarity analyses are based on flight histories of specific components and subsystems. In any event, the SE principle of complete verification is paramount. This chapter will show how 100 percent verification is accomplished for aircraft systems. It also shows how traditional aircraft processes, such as certification, are incorporated into the SE verification process.

This chapter also explains the difference between verification and validation and how requirements validation differs from product validation.

11.1 The Verification Matrix

All verification starts with the verification matrix. This matrix is a standard part of the specification format as shown below. Basically, the verification matrix assigns a verification method to each requirement, thus fulfilling the SE goal of 100 percent verification.

For the certification data package, the verification matrix has specific required contents, as follows:

1	Requirement.	4	Verification method(s) applied.
2	Associated function.	5	Verification conclusion (pass or fail).
3	Development assurance level.	6	Verification coverage summary.

11.2 Traditional SE Verification

Traditionally, SE has recognized four categories of verification: test, demonstration, analysis, and examination. Examination is sometimes called inspection. Some sources include simulation and similarity as separate categories. We have included these as types of analysis.

Test

A test is a type of verification which requires instrumentation. These tests can be, for example, pressure tests, wind tunnel tests, or flight tests. There is sometimes confusion between development tests and verification tests. A development test is a test conducted to reach a desired design and to satisfy requirements. Trade-offs can be conducted during development tests. Wind tunnel tests can be development tests. They can also be verification tests. In fact, the same test can be both a development test and a verification test. The difference is that only the test which represents the final solution will be the verification test. It is the test that will verify (or be part of the verifications), for example, the requirement to provide a given lift capability. All the other test points qualify as development tests.

Testing is an essential element in the certification process. Testing assures that every aircraft element performs the function it was intended to perform, performs it to the expected level of performance, and does not perform functions it was *not* intended to perform.

Analysis

Analysis is any kind of mathematical, computational, or logical task performed to verify a requirement which cannot be verified in any other manner. In addition, analysis is used as an initial type verification to assure that the aircraft meets the requirements early in the development phase. The results of the analysis will be confirmed by actual flight tests later in the program. Analysis can also be used in conjunction with testing to extend the envelope of the test results. For example, if the aircraft is tested at certain points in the flight envelope, computer simulations can be used to predict the aircraft behavior in regimes beyond and in between the actual flight data points.

Analysis, from the certification point of view, does not include similarity, which is discussed in Section 11.2.

Simulation

Verification by simulation is a type of analysis using computers. Simulation is used extensively in the commercial aircraft industry. The flight handling characteristics and aircraft performance characteristics can be simulated and used to verify performance requirements long before they are tested in flight. Simulation can also be used in ergonomics to simulate the movement of the human body when interacting with aircraft components. Analysis by computer simulation is also a valid method of verification for certification. Typical simulation types include finite element analysis (FEA) to analyze structures. Another type of simulation is computational fluid dynamics (CFD) which analyzes air flow around the aircraft.

In-service data and similarity

Analysis by similarity is a frequently used method of verification. It is based on the assumption that another component which has met *the same performance requirements* and operated *in the same environment* will meet its own performance requirements. In the aircraft industry this similarity is based largely on *in-service data*, that is, on past experience of another component in another aircraft. Great care should be taken, however, to assure that the operating conditions of the comparison aircraft are equivalent to the aircraft being designed. In fact, no two aircraft are identical and no two operating conditions are identical. One aircraft may have operated in extremely cold climates, while the other may have operated in very hot climates. When the supplier wishes to use an existing component as a solution, a Letter of Similarity should be submitted to the manufacturer justifying the past use as a verification of the requirements. In-service data is also a valid method of verification for certification.

System safety assessment (SSA)

Analysis also includes the SSA, which is one of the elements required by the certification process described in Section 10.2. SSA is used to verify that the system meets the safety requirements.

Demonstration

Demonstrations are similar to tests but do not require any instrumentation of any sophistication. Typical demonstrations might include the functioning of an emergency alarm, for example. The human actions simulated in Section 11.2 during concept development can now be demonstrated with real components and real humans.

Examination

Examinations are the easiest type of verification. They are simply a visual confirmation that a requirement has been met. There can be both drawing examinations, to determine that the required equipment has been included in the drawing, and hardware examinations, to confirm that a piece of equipment has been installed on the aircraft. This category is also called inspection.

In the certification process, inspection and review is an official verification category. In addition, certification calls for reviews to assure that the product complies with the requirements.

11.3 Verification of Regulatory Requirements

Certification calls for a complete verification program and for the manufacturer to document the program of verification in the certification plan shown in Table 10.1. Of particular interest are the regulatory requirements and the specified test conditions found in the FARs. But, from an SE point of view, verification of the regulatory requirements is only part of the total picture.

11.4 Verification of Customer Requirements

In the traditional world, verification of customer requirements is a most often neglected task. This neglect is most often seen with respect to buyer-furnished equipment (BFE). If we do not adhere to the general principle that all requirements should be verified, then there is a risk that the customer equipment will not perform in the aircraft environment as it was expected to or that a safety hazard may be introduced.

11.5 Verification Sequence

Verification sequence is extremely important. If you wait until the product is built, that may be too late. Early verification using simulation and analysis will reduce this risk.

Verification takes place throughout the development process. Analyses and development tests done during the design process become the first steps in verification. Drawing examinations done for CDR are part of verification. Then qualification testing done at supplier facilities, followed by integration testing, and finally flight testing, are all part of the verification process.

Verification and system integration are inextricably linked. The manufacturer is required to develop and document a system integration plan in which the order of mating and testing is done. It is a bottom-up process: the lowest-level components of the aircraft hierarchy are mated and tested first, followed by higher-level integration and testing.

11.6   System Validation

When all requirements have been verified and the aircraft has been assured of meeting its mission objectives, this state is called *system validation*. System validation requires involvement by the customer because it is the customer who validates the system. This can be done either with flight tests or by inspecting the verification documents that have been developed.

There is an implication in the term *system validation* that the documented requirements may not satisfy all of the customer needs. If this were not true, there would be no need for system validation. If the customer is not satisfied with the aircraft as built and as flown, then further effort to satisfy these needs will be the subject of negotiations between the customer and the manufacturer.

System validation should not be confused with requirements validation as explained in Chapter 4. Requirements validation, as explained in ARP 4754A (2010, p. 13), is the assurance that the requirements are correct and complete. Although the same word is used to describe these two concepts, they are, in fact, quite different concepts.

11.7   Qualification

According to Kossiakoff and Sweet (2003, p. 452), qualification is the process of proving that a product or process meets all of its requirements. Hence it is the final decision that the aircraft is ready to be delivered and fly. There is a strong emphasis in qualification on environmental testing. That is to say, qualification assures that the aircraft has been tested in all of the environments it will experience in service.

Systems Engineering Management and Control

SE without good management is like an orchestra without a conductor and all the musicians are playing to a different sheet of music. In most cases, the introduction of SE into a commercial organization will result in significant changes to the management operations of a company. This section provides some guidelines to make the process of introduction easier and also some of the basic features of SE management which are essential.

SE management refers to those activities which are programmatic in nature. Control, on the other hand, refers to those activities which are conducted to assure the integrity of the SE process and the quality of the product. These include configuration management, risk management, data management, SE task scheduling, technical performance measures (TPMs), and design reviews. SE management includes the implementation of an integrated product development (IPD) program.

In spite of the fact that we have left SE management to the twelfth chapter in this book, many managers consider SE management to be *the* SE process. In fact, one major book by Kossiakoff and Sweet (2003) considers SE to be a subset of management. Therefore, its position in the book does not reflect an inferior position of importance. It is very important. Furthermore, it is recommended that within the IPD concept, integrated product team (IPT) leaders should be thoroughly familiar with the SE process and not rely on SE staff members.

12.1   Management Responsibilities

Strong management is essential for the execution of an SE program. In addition, in the commercial world management may have to do things a little differently. Following are a few guidelines for the manager:

SE planning

Traditionally, the SE manager is responsible for preparing an SE management plan (SEMP) to describe all the SE activities on a program. However, in the interest of creating fewer documents, in the commercial world the SEMP can be included in the program plan.

In addition, the planning must include the certification and the development plans shown in Table 10.1 required for certification.

Integrated Product Development (IPD) phased schedule

The IPD phased schedule, described below in Section 12.3, recommends that all program phases be scheduled and completed before a new phase can begin. Strong management is needed to make this schedule happen.

Personnel staffing

An SE program recommends up-front staffing with the expectation of down-stream savings. SE management should plan for and acquire this staffing to make it work.

Work order

As we have seen in this book, SE recommends that tasks be done in a different order. It recommends, for example, that customer requirements, functional analyses, and requirements analyses be done *before* design begins. This order means that SE should begin *really early* and that drawing development should not begin before the other steps are complete. Otherwise, the chances are high that these tasks will have to be redone when the requirements are finished.

Program integration

The SE manager should assure that the SE tasks are integrated into the balance of the program. The manager should assure, for example, that the designers are working to the developed requirements and that the identified SE trade-offs are performed and implemented. The SE manager should assure that interface meetings between all interface parties are scheduled and conducted as discussed in Section 6.6.

Design reviews

As we will also see below, design reviews are critical elements in SE. SE-driven design reviews require planning and organizing. The manager should make sure that all the data have been developed *and reviewed* before each review. The manager should ensure that all personnel are present. For example, representatives from manufacturing and product support are essential as discussed below in Section 12.4. Most important, the *customers and suppliers* are necessary attendees at SE design reviews. Finally, the manager is responsible for assuring that the design review is closed out and that action items are complete.

Also, as will be discussed in Section 12.4, design reviews are one aspect of SE that the commercial aircraft company will find most demanding in terms of time and money. The question is: Can these reviews be conducted in more timely and cost-effective way without increasing risk? This question will be addressed in that section.

Deliverables

The SE manager should assure that all SE deliverables, such as specifications and ICDs, are complete and on schedule.

Manage Technical Performance Measures (TPMs)

SE management should manage the TPMs as described below. Many managers consider TPMs to be most useful SE tool from a management point of view.

SE principles

The SE manager has the sole responsibility of assuring adherence to the SE principles outlined in this book, the two most important being: first, the design should incorporate *all* requirements, and secondly, *all* requirements should be verified. Additionally, management should assure that requirements are identified according to the holistic approach described in Section 4.1. Finally, of all SE principles, risk is the principle for which managers should take total ownership.

Configuration management supervision

Although the SE manager does not have direct control of configuration management described in Section 12.11, this is one area requiring constant oversight and reporting. It is important that *all* engineers working on a given project are working to the same configuration baseline. Within the IPD framework described in Section 12.3, the IPT has responsibility for the configuration management of its own segment of the aircraft.

Project redirection

One of the most important roles of the SE manager is to make sure all SE processes are in place and being followed during periods of project redirection. For example, a crucial time is when new requirements appear (from the customer, for example) after preliminary design review (PDR). We will see in Section 12.4 that PDR is the milestone at which all requirements are frozen. Hence, when new requirements appear after PDR, the SE manager should assure that all SE steps are taken to assure the proper incorporation of those requirements.

Supervision of risk management

One area requiring close management scrutiny is risk management (see Chapter 15). Risk management will identify many areas of program cost, schedule, and technical risk and give the manager the information with which to make educated decisions. To initiate a risk management activity, the SE manager needs to place full responsibility for the risk analysis in the hands of the program systems engineers and to assure that a non-advocate risk assessment team is in place.

Verification supervision

In order to implement the SE principle of complete verification, the SE manager should assign a single person or organization to coordinate all verification. The scope of this task is broader than design or test; hence, this person would normally be part of the SE group.

12.2   The Chief Systems Engineer (CSE)

One issue which has been debated in government-oriented organizations for years and is new to commercial organizations is where the person responsible for managing the SE activities should fit into the program organization. This person is normally called the Chief Systems Engineer (CSE). Ideally the program manager and/or the Chief Design Engineer (CDE) would be so familiar with SE principles that a CSE would not be necessary. If that is not the case, then where should the CSE fit? The CSE could either be parallel to or over the CDE.

The global scope of SE as outlined in this book argues that the proper position for the CSE is as the assistant program manager and hence over the CDE. This concept does not in any way diminish the importance of the CDE but rather places the CSE in a position to execute the SE activities as needed. We will see in Section 12.3 how the roles of program management and SE merge within the framework of IPD.

12.3   Integrated Product Development (IPD)

The basic idea behind IPD is that a system can be developed more efficiently and faster if the various aspects are done *together* rather than separately and in parallel. We can view IPD as the ideal management process within which SE is accomplished. There are many IPD principles. But following are the key ones, especially as they affect aircraft development.

Multifunctional teams

Multifunctional teams are especially important in aircraft development. This importance derives from the fact that aircraft development involves such a large number of technologically diverse and geographically dispersed technologies and disciplines. Multifunctional teams play the role of bringing these groups together. Hence, working together in the same room are such diverse groups as safety, human factors, maintainability, manufacturing, and aerodynamics. The presence of all these groups is recommended at design reviews, as described in Section 12.4, and in other technical coordination meetings. IPD also encourages the participation of customer representatives and suppliers.

Another important function of multifunctional teams is to reduce the possibility of hidden interactions that may occur between the components of different design disciplines. As discussed in Section 16.4 these types of interactions have been the root cause of accidents in many different domains. Multifunctional teams are one way of implementing the *reduce hidden interactions* rule discussed in that section.

Phased development

The key principle in phased development is that no new development phase should be initiated until the previous one is complete. The principal milestones of each phase are the design reviews described in Section 12.4. The value of this principle is that it removes much of the development risk inherent in a process in which major modifications may result from prematurely initiating tasks which are dependent on design aspects which have not yet been frozen.

Integrated Product Teams (IPTs)

IPTs are one area that has significantly altered the way large systems are developed and built. What IPT means is that parts of the aircraft are developed as *products*, such as the wing, nose, empennage, and so on, rather than as functional areas, such as avionics, mechanical, and so on. For example, the nose organization would all work together, including the electrical, mechanical, avionics, and other personnel. In addition, manufacturing and other personnel would be members of the nose IPT. This arrangement helps communications enormously.

A key aspect of the IPT is the principle of requirements ownership. That is, the IPT owns all aspects of requirements. For example, the IPT is responsible for implementing requirements allocated from higher-level IPTs. These might include, for example, weight or dispatch reliability. Secondly, the IPT is responsible for developing and allocating requirements within its own domain. Finally, the IPT is responsible for every aspect of implementing its requirements. This includes design, procurement, fabrication, installation, and verification.

Another aspect is that each aircraft segment (the product) would be integrated *and tested* completely before being joined with the other segments. This methodology is completely compatible with the bottom-up integration principle. Another way to put it is that the nose (and wing, and so on) would be *stuffed*: that is all the electrical cabling, environmental ducts and so on would be in the nose before it is joined with the fuselage.

The IPT philosophy creates a problem, not insurmountable, for the SE process. The problem is that the *distributed* subsystems (electrical, hydraulics, and so on) are now resident in the various aircraft segments. Thus, the systems engineer should treat all of these subsystems as if they were separate subsystems in each segment. For example, there would be the nose electrical subsystem, the fuselage electrical subsystem, and so on. Each of these subsystems will interface with the other subsystems, thus creating additional interfaces which would not have existed as external interfaces if the whole electrical system had been treated as a single subsystem. In addition, each electrical subsystem's performance requirements will have to be separately specified and verified.

Another key issue with respect to IPTs is how SE should fit into them. Traditionally, SE was considered a staff function. The systems engineer would advise the IPT leader on various aspects of the SE process. Current SE thinking has shifted away from the staff function approach and towards a philosophy that the IPT leader should be a person thoroughly educated in SE methodology and outlook. This philosophy puts a strong responsibility on the training program.

12.4 Design Reviews

Quality design reviews are undoubtedly the most important aspect of SE management. Design reviews are where we establish that all requirements have been identified, that we have a design that meets the requirements, and that the design has been verified. We saw above that design reviews are critical to the IPD phased development and that the development program cannot proceed until each design review has been completed *and is successful*: that is, all the criteria have been met. We also saw that, using the principles of IPD, the design reviews are where the *entire development community*, from marketing, to manufacturing, to engineering, to supplier management, to the customer, gathers to put its stamp of approval on the design.

A key aspect of design reviews, from the aircraft point of view, is that *aircraft-level* reviews are an integral part of the process. A logical question is then: Is an aircraft-level review recommended for all aircraft changes. The answer is no. So what are some guidelines for conducting aircraft-level reviews? Following are a few. Aircraft-level reviews should be held for:

1. New or derivative aircraft designs.

2. Collections of changes, the accumulation of which will affect aircraft-level parameters.
3. Any change affecting aircraft-level parameters.

The second logical question is: What is an aircraft-level parameter? There are many parameters measured at the aircraft level. The most obvious one is weight. Any component weight will, by definition, change the weight of the entire aircraft. If the weight change is significant, then center of gravity (c.g.) and other mass properties may also change. Another aircraft-level parameter is dispatch reliability. Any subsystem change affecting the aircraft handling capability will be a candidate for aircraft-level review.

Another important aspect of design reviews is that they are a critical requirement for certification.

Tailoring the design review process

The challenge of design review tailoring is to make the reviews both faster and less labor intensive and at the same time rigorous. But before we examine some of the options, let's look at some of the requirements that cannot be sacrificed:

First, all necessary personnel should participate, and their participation must be substantive and not just a formality. They must actually examine the material, make a professionally educated judgment and not just sign a form. In addition, the phrase "all necessary personnel" must include not just the immediate designers but members of specialty disciplines as well.

The above discussion raises the question of whom to include and whom to excuse. This question should be answered very broadly. That is, invite anyone for whom there may be a remote possibility of involvement. For example, any component with an exposure to outside air should trigger the attendance of aerodynamicists. As another example, any component emitting EMI (electromagnetic interference) or vulnerable to EMI should trigger the attendance of an EMI specialist.

Secondly, it is a common misperception that attendees at design reviews are invited to find mistakes in the material or to criticize it. On the contrary, it is the duty of the attendees to examine the material well *in advance* and provide that feedback to the designers so that they can make those corrections. So signing the design review approval forms will indeed be a formality since all comments will have been treated in advance.

One aspect of design reviews is particularly subject to adaptation. That is any aspect that has to do with requirements. If the SE organization has thoroughly reviewed the requirements according to the guidance of Chapter 4, much time can be saved. This is particularly true of the review of supplier specifications. The SE organization will assure, at a minimum, that all requirements satisfy the criteria of Section 4.10.

The SE organization will have to do more than just determine that the requirements are written in the proper format. They will have to determine that requirements have been allocated or derived correctly. In discussions with the design organization the SE organization will have to determine that the requirements comply with engineering design principles of the discipline at hand. In short, if the SE organization performs this task, there may be very little else to do and the formal design review can be eliminated entirely.

Finally, it cannot be overemphasized that accomplishing the two requirements above will require strong action by the program manager since these actions are not commonly adhered to. Now on to some options:

The electronic review option
In the days of electronic communications this would seem to be the logical option. In this option all material (specifications, drawing, and so forth) would be sent to the attendees electronically. Attendees could comment on the material and then sign off on an approval when that is required. Responses could be sent to all attendees for their consideration. At the risk of repetition it must be remembered that all attendees who have been selected as necessary reviewers must scrutinize the material from the point of view of their expertise. In this option the review organizer must specify exactly what materials and specific pages the attendees must review.

The video review option
This option is essentially the same as a live face-to-face meeting except that the attendees use their computers to see and hear the attendees. They still must receive the material electronically as above and respond to it.

The wall walk option
This is an option that is practiced in organizations today and is particularly useful when drawings are to be reviewed, such as for the critical design review (CDR) described below. In this option the drawings are attached to a wall in a large room in which the attendees can examine the drawings and mark them up when they apply to their area of expertise. For example, if a drawing contains both an electrical conduit attached to a structural member, engineers from both the electrical and structures department must review it and mark it up as necessary.

The following paragraphs summarize the individual design reviews.

System Requirements Review (SRR)

The SRR is the first major design review, and a very important one at that. The focus of the SRR is on the top-level requirement and not on the design. Top-level requirements are those requirements which are not derived: that is, they do not depend on design solutions as discussed in Section 4.2. The purpose of the SRR is to verify that all the top-level requirements are correct: that is, that they meet with customer approval. That is why the presence of the customer is important. At the aircraft level, if there are many potential customers, several SRRs may be recommended. Or secondarily, *surrogate customers* may be recommended, that is, members of the marketing department who have communicated with the various customers and can speak for them. In this case it is important to summarize the SRR for the various customers to obtain their concurrence. In the case of subsystem-level projects, there will normally be a single customer who can attend. Subsystem SRRs should be held after the aircraft-level SRR.

The focal point of the SRR will be the *mission statement* discussed in Section 4.3. This mission statement will capture both the key customer requirements and also describe the customer operational objectives and the environment in which the system must operate. The mission statement is a key part of the system specification, as described below.

Another function of the SRR is to present to the customer those *assumed* requirements that have been developed throughout the requirements development discussed in Section 4.4. In addition, the SRR should present to customers the capabilities of the buyer furnished equipment (BFE) they have requested. It is a common error to assume that the requirements for the BFE do not have to be examined "because the customer asked for it." Failure to examine the capabilities of the BFE and report these capabilities to the customer may result in either or both a performance shortfall (of the item) or an unexamined safety hazard.

Although the SRR will not present any solutions, it is permissible to show a preliminary concept. In addition, a *system architecture*, as shown in Figure 2.1 will be presented. The system architecture is the hierarchy into which requirements will be allocated as discussed in Section 4.7.

System Design Review (SDR)

The SDR will be the first review of a concept which meets the top-level requirements. In addition, the SDR will show requirements which have been flowed down to a level *one level below the top level* in accordance with the principle of top-down allocation discussed in Section 4.7. Another aspect of the SDR is that it is the point at which any major trade-offs are initiated. The SDR identifies these trade-offs. Subsystem SDRs should be held after the aircraft-level SDR.

Preliminary Design Review (PDR)

The PDR has, perhaps, a misleading name because there is nothing very preliminary about it. In fact, the requirements should be defined to the lowest level of the aircraft hierarchy. In addition, the PDR should define a design to meet all those requirements. The main function of the PDR is to initiate the detail drawing process. The PDR marks the end of the front-end SE activity. Although SE continues for the life of the development process, the tasks to follow are primarily in verification. The PDR also marks the end of major customer participation. If the customer creates additional requirements after the PDR, then program management will have to alter the program schedule to accommodate them. In short, program management will have to reinitiate the requirements process. Subsystem SRRs should be held before the aircraft-level SRR.

Critical Design Review (CDR)

The CDR is not really an SE review. It is a real *design* review. That is, the purpose of the CDR is to make sure the detail drawings agree with the concept completed at the PDR. Once CDR is complete, engineering can release drawings to manufacturing, not before. If engineering does release drawings before the CDR, development risk will result.

The exception to the above rule occurs with the release of drawing for long-lead-time items. These are items whose development time exceeds the development time allowed in the program schedule. However, the release of drawings for long-lead-time items does not relieve the development risk. In this case program management has agreed to *accept* the development risk.

Subsystem CDRs should be held before the aircraft-level CDR. However, drawing release should not occur until completion of the aircraft-level CDR.

Physical Configuration Audit (PCA)

PCAs answer the question: Was the aircraft built in accordance with the drawings reviewed at the CDR? In addition, the PCA fulfills the requirements of the audit requirements of certification.

Functional Configuration Audit (FCA)

The purpose of the FCA is to verify the functionality of subsystems as they are installed on the aircraft. These reviews are also part of the reviews designed to accomplish certification.

System Verification Review (SVR)

The system verification review is the most important review prior to the delivery of the aircraft. The SVR verifies that the aircraft *as built* meets the requirements, both performance and constraints, developed in Chapter 4, Requirements. The SVR is also a key step towards certification.

First Flight Review (FFR)

This review ensures that the first test aircraft is ready for flight. All verifications should have been done to that date, and all extra requirements, such as for instrumentation, should be met. The FRR is also called the first flight readiness review (FFRR).

Engineering Safety Review (ESR)

ARP 4754A (2010, p. 104) also recommends an engineering safety review. The purpose of this review is to assure that subsystems were built in the correct configuration without flaws or errors that may affect the safety of the aircraft.

12.5   Documentation

Documentation is the life blood of SE. Although in the commercial world there is less documentation than, say, in the military world, documentation is a necessary element for maintaining records of requirements, solutions, and verification. As we point out in Section 12.5, the word *documentation* can be used in its broadest sense to include electronic records, so that no actual paper is required.

Specifications

In the commercial world the term *specification* generally refers only to those documents which reflect agreements between the manufacturer and the customer or between the manufacturer and a supplier. Rarely are internal specifications produced to reflect the requirements for internally defined segments or subsystems. However, with the advent of SE, this internal allocation, either through a specification or another medium, such as a requirements data base, is becoming both essential and more common.

Internal specifications

In an SE environment, internal specifications will maintain a performance and constraint focus. Unlike customer specifications, the internal specifications will evolve as new requirements are refined and derived. The usual way to handle this evolution is to use TBDs (to be determined requirements) as placeholders for values of requirements to be determined at a later date. The presence of a TBD in a specification implies a task (by the manufacturer or the supplier) to determine the value. Therefore, there should be a clear linkage between the specifications and the statement of work (SOW), discussed below.

There is no standard format for internal commercial specifications. However, the Institute of Electrical and Electronic Engineering (IEEE) has published such a format, as documented in IEEE 1233 (1996). A much more common specification format is the DoD-developed MIL-STD-961D (1995) often called the *six-part* format, which contains many topics of interest only to the military community. Appendix 2 provides a version of the MIL-STD-961D format, slightly modified for commercial application,

Customer specifications

A customer specification is a contractual document rather than an engineering document. The customer specification has a total configuration focus and will reflect the aircraft *as it will be delivered to the customer*. It is not a good practice to make a single document fulfill both roles.

Procurement specifications

In the SE world, procurement specifications will cease to exist as separate documents since internal specifications will contain all the material needed for suppliers to design their components.

A "paperless" SE process

One of the greatest obstacles the systems engineer will encounter is the reluctance by design engineers to produce more paper. And they are right. Who needs more paper? But on the other hand, accurate recording of requirements is essential to the SE process. What can we do? There are several possibilities. First, the automated SE tool can organize and print out the specifications. Secondly, the systems engineer can conduct brainstorming sessions and keep all the records. The design engineer only provides verbal information.

Interface documents

As described in Chapter 6, Interfaces, the primary interface document is the Interface Control Drawing (ICD). The ICD will contain all the important information about interfaces, including the functional interfaces, a sometimes neglected topic in the commercial world.

Statement of Work (SOW)

In the commercial world, the SOW is often a mixed bag of requirements, solutions, and tasks. In fact, specifications often contain solutions and tasks as well. It is a basic principle of SE to make a clear distinction among these three media. The *only* function of a SOW is to describe tasks, either for the aircraft manufacturer or for a supplier. The risks of not making this distinction are that, first, if requirements are put into the SOW, they may not be incorporated into the design or verified; and secondly, if tasks are put into the specifications, which are technical requirements documents, they will most likely be ignored and not performed.

Baseline Concept Document (BCD)

The purpose of the BCD is to control the *physical configuration* of the aircraft. No performance data should be included in this document. The BCD will evolve during product development to capture *and control* the configuration as it is known at any time.

In addition to the description of the aircraft, another important part of the BCD is a description of the operational concept. Rather than describing how the aircraft looks, the operational concept describes how the aircraft will be used. For example, an aircraft might be designed to be used at airports with minimal support equipment.

12.6 Automated Requirements Tools

Automated requirements tools are a key method for systems engineers to develop and track functions, requirements, and solutions. As such, their support and control by SE management is valuable. We saw above that requirements documents, such as specifications, need not be hard copy items but can be electronic. Automated tools are a key way to accomplish this electronic data management. Appendix 3 provides a summary of automated tool value and characteristics.

12.7 Technical Performance Measurement (TPM)

TPM is a management tool for tracking and managing requirements compliance of selected parameters. TPM continuously verifies the degree of anticipated and actual achievement for technical parameters. TPM confirms progress and identifies deficiencies that might jeopardize meeting a system requirement. TPM assesses values which are outside established tolerances and thus indicate a need for evaluation and corrective action.

TPM is the key tool for managing the top-down allocated requirements described in Section 4.7. With TPM the program manager has total visibility of which subsystems may not be meeting their allocated values of, for example, weight or dispatch reliability. With this information, the program manager has three options: The first option is to ascertain whether the subsystem with an excess weight, for example, can take measures to meet the weight requirement. The other option is to reallocate the weights. If this option is taken, some other subsystem will have to live with an even more stringent weight requirement. The third option is, of course, the least desirable option, namely, that the entire aircraft will weigh more than expected. Even if this is the only possible outcome, management will be fully aware that all options have been examined.

12.8   Software Management

Although SE views software as a subsystem, like any other subsystem, management guidelines exist which govern the management of software, especially with respect to the certification. Many tasks described in Table 10.2 pertain to software development, integration, and testing.

12.9   Supplier Management

Supplier management, also called *procurement,* is one of the most critical aspects of SE management in the commercial aircraft industry because many aircraft components are manufactured by suppliers.

Supplier requirements

Requirements should be provided to the supplier in a thorough and rigorous manner. Hence, the rules for requirements development and documentation become even more important. In the traditional environment, the manufacturer develops all requirements and most often provides the supplier with only build-to specifications. In the SE environment, the supplier is responsible for the *performance* of the product, not just the physical configuration. However, the most important principle for supplier management in the SE environment is that the aircraft manufacturer owns all requirements and is responsible for assuring that the supplier's product meets those requirements.

That is, it is not good management practice to *assume* anything. The manufacturer should, first, provide the requirements to the supplier as described in the following sections. Secondly, the manufacturer should require that the supplier provide evidence that the product will meet the requirements. Finally, the supplier should provide evidence that the product, as built, meets those requirements, that is, verifies the requirements in accordance with the principles of Chapter 11.

Requirements flow down to the supplier

In the SE environment the supplier is given only the subsystem-level requirements (that is, one level above the supplier's product). Often the subsystem-level requirements include TBDs in cases where the system-level requirements are dependent on supplier input. In addition, these requirements are in the form of performance requirements, not solutions. As part of the supplier's contract, the supplier helps develop both system-level requirements and derived supplier product requirements. The net result of this process is that the supplier can optimize the supplier product requirements and can propose solutions perhaps more cost effective than those envisioned by the aircraft manufacturer. In summary, requirements can be categorized as follows:

1. Requirements developed by the aircraft manufacturer.
2. Requirements developed jointly by the aircraft manufacturer and the suppliers.
3. Requirements which are solely the responsibility of the supplier.

The key principle is that the manufacturer *owns* all requirements, retains those which the SE process dictates, and delegates to the supplier only those below the subsystem level, that is, below the manufacturer–supplier interface. The reason for this ownership principle is that the supplier does not have the *span of knowledge* to develop the subsystem-level requirements, as discussed below.

Span of knowledge

One of the common mistakes often made by engineers not trained in the rigors of SE is the violation of the *span of knowledge* principle. This mistake is particularly damaging when dealing with suppliers. This principle states, in summary, that the information contained in the requirements for any element should be the complete information and the only information required to define the requirement. As an example, if the operation of any subsystem needs to know the altitude of the aircraft, then that information should be supplied to the subsystem in the manner that it matters to the subsystem, for example, in terms of pressure or temperature.

The supplier in the Integrated Product Development (IPD) process

Since the SE process can be viewed as a subset of the IPD process, the supplier will be a member of the IPD teams throughout the product development.

The supplier in the synthesis process

Except for the subsystem-level requirements, the constraints, and the interfaces provided by the manufacturer, suppliers will have a free hand to develop and build their own design. In this way a clear division of responsibility between the manufacturer and the supplier can be developed and the best design can be built.

The supplier in the verification process

If the supplier is responsible for the performance of the product, as we said in Section 12.9, then it follows that the supplier is an integral part of the verification process. The supplier's role in the verification process begins with the verification matrix shown in Section 11.1. The verification matrix will specify for *every* requirement the responsibility for the verification of that requirement. Each verification method (test, demonstration, analysis, inspection) will be incorporated into the supplier's test plan.

Supplier cost control

It is a key function of supplier management to assure the lowest cost of the procured items for the aircraft. Item cost affects both the cost of the aircraft as well as the cost of spares discussed in Section 5.7. To this end, supplier management should assure competitive bidding among suppliers and avoid sole-source acquisition. This factor should, of course, be balanced against the desire to reduce the lead time in the development process.

We saw in Section 6.1 that one method of encouraging greater competition among suppliers is by requiring common interfaces through the implementation of SAE standard AS4893 (1996).

12.10   Configuration Management

Configuration management is part of the certification plan (Table 10.1) and one of the pieces of certification data in the configuration management plan. The importance of configuration management in the SE process is that it assures the continuity and the integrity of the results of the synthesis process.

Configuration management consists of configuration identification, control, verification, and accounting. Configuration identification is the baseline documentation itself. Configuration control is the formal process of controlling changes to the baseline. Configuration verification is the process of assuring that the system meets the intent of the customer. This process is conducted through reviews and audits. Configuration accounting maintains configuration data and tracks the status of the configuration.

Configuration management controls both the configuration of the aircraft as well as the data required to define the aircraft. These data include all the certification data shown in Table 10.1. The configuration management plan includes the method to show that the objectives of the configuration process are satisfied.

A key ingredient in configuration management is the configuration index. A configuration index is a catalogue of the physical elements which comprise the aircraft and its subsystems. The specification tree shown in Figure 2.1 shows a typical hierarchical structure on which a configuration index would be based. The primary index system used in the aircraft industry is the Air Transport Association (ATA) Specification 100 index. This index can be modified or adapted to the needs of a particular aircraft application. The certification data set only requires that a single index be employed in the development of an aircraft. When using the ATA index in the SE context, it is important to group the elements of the index so that they are compatible with the functions as defined in Chapter 3. The configuration index will contain the identification of each element, associated software, interconnection of elements, interfaces, and safety-related procedures and limitations.

12.11   Integration Planning

System integration is the task of assuring that all items work together individually and collectively as a group or as a whole aircraft. A system is built by taking the lowest-level components and putting them together one level at a time. The cornerstone of system integration is that it is a bottom-up process. Between each level's integration, it is necessary to test the lower levels to make sure they work together. The verification process discussed in Chapter 11 will show the verification at all levels of aircraft integration. If lower-level components subsystems are tested first, problems can be uncovered before higher-level assemblies are integrated and tested. The system integration will determine the verification sequence of the system build-up.

As an example, take the verification and installation of the environmental control system (ECS). Before delivery of the product, the supplier will conduct verification (test, demonstration, analysis, or inspection) on each component made by that supplier. For example, the supplier will verify the performance of the air supply unit under prescribed electrical loads and input air sources. However, this verification will not assure that the ECS meets the aircraft-level requirements, namely, to deliver a specified air flow from all ducts and to maintain a given air temperature in the cabin. To verify the aircraft-level requirements, testing should be conducted at the aircraft level, with *passengers on board*. The importance of this principle is that only at the aircraft level can the requirements be verified under the operational conditions.

Hardware and software integration documents the entire process of integration of an aircraft and its assemblies. This process may include breadboards, prototypes, computer emulations, and laboratory or flight-worthy items. Documentation of the integration process is a certification requirement as shown in Table 10.1.

Another important part of integration planning is interface management discussed in Section 6.5. It is a role of program management to assure that the different parties to interfaces are brought together to agree on interface responsibilities and to assure that interface requirements (both functional and physical) are met.

13

Adapting Systems Engineering to the Commercial Aircraft Domain

A theme of this book is *adaptation*, that is, how should an organization adapt the SE process to make it both effective and affordable. Hence this book can be seen as more of a *how* and a *why* book, rather than a *what* book. The first part focuses on the adaptation process and the importance of risk in that adaptation. The second part shows how an existing organization can be adapted to incorporate SE.

The organization itself can be viewed as a system to which the principles of SE would apply. In fact, the *NASA Systems Engineering Handbook* (1995) describes total quality management (TQM) as "the application of systems engineering to the work environment." The NASA handbook also points to other similarities between TQM and SE, such as the emphasis on customer satisfaction.

13.1 Adapting the Process

One of the principal themes of this book is that SE does not have to be practiced *in its entirety* to be effective. So the question is: If SE is not practiced in its entirety, how does the program decide what not to do? The answer lies in two rules: (1) perform what has to be done to achieve the objective on any project, and (2) delete what has been thoughtfully considered to be low risk. Figure 13.1 illustrates these rules.

This figure is admittedly notional, but it makes an important point, namely, that the more SE you perform, the lower the risk. However, usually more SE comes with two consequences: time and money. So the question is: How do you know what to keep and what to delete?

Experience will tell you what to keep. Most projects boil down to one or two really important requirements. For lightning projects, the question is how much current can flow through any section of the structure without creating a spark to ignite fuel. For brake design the question is how much energy the brake pads have to absorb to stop the aircraft on the runway. For a project involving the replacement or incorporation of a new software module, it is absolutely essential that the interfaces with surrounding subsystems be correct. Chapter 9 describes the typical parameters for various subsystems.

Deleting steps is a much harder process. The important point is that steps should not be deleted without consulting experts in all disciplines. For example, any step critical to the safety of the aircraft should not be deleted. Beyond

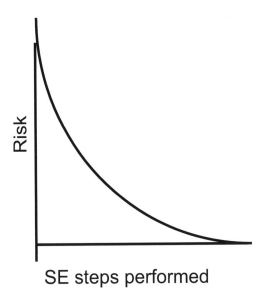

Figure 13.1   The SE adaptation diagram

safety, every project should be examined to determine what disciplines would be affected, for example, human factors, reliability, and so forth. These deletions can be determined with a brief review of the project plan by all disciplines. If this review is conscientious and thoughtful, these experts can be excused from further participation.

Chapter 4, Requirements, presents a prime example of the application of the concept shown in Figure 13.1. The question at hand here is: what requirements can be deleted without incurring excessive risk? Chapter 4 shows how the number of constraints can be so overwhelming that the inclusion of all of them would put a great burden on the project. So Chapter 4 suggests that key people can make the decisions on which requirements to delete. These decisions will be based on risk.

Following are a set of guidelines for adapting SE to the commercial aviation domain. This list is summarized from a paper by Jackson (1996).

Tribal knowledge

All organizations exist to some degree on *tribal knowledge*. It is not the role of SE to replace that tribal knowledge but to complement it. As an example, let's say that this organization has been procuring pitot tubes from the same supplier for many years and that the pitot tubes have been of extraordinary quality. The pitot tubes have lasted the entire life of the aircraft without replacement and never fail; after all pitot tubes are an essential component on an aircraft. The lesson is simple: stick with this supplier.

Existing processes

If the commercial organization has been in business for a long time, there may be many processes in existence. These processes may cover a variety of topics: for example, customer relations, procurement, inspection, testing, and so forth. In fact, many organizations may have processes that are recognized SE processes, such as configuration management, integration, or integrated product development (IPD). But these organizations may not implement SE as a whole. Thus, it is the best policy to make the SE process complement the existing processes. It may be necessary to change the existing processes. But it is never wise to ignore the existing processes.

Existing SE

SE may already exist in the commercial aircraft company, either because it is a result of good engineering judgment or because it has been mandated by other outside agencies. SE may be reflected in current documentation. Existing documentation may include change requests, engineering work orders (EWOs), specifications (both customer and supplier), interface documents, supplier packages, test plans, test reports, and many other documents. Mission statements (partial or complete) may appear in the change requests, EWOs, and other documents. Likewise, these same documents may provide partial or complete functional analyses. Requirements will appear in specifications (although specifications in the commercial world tend to emphasize the product design characteristics). Test plans and reports will contain much needed verification information.

We saw before Chapter 10 that the Society of Automotive Engineers (SAE) and the Federal Aviation Agency (FAA) have jointly developed a guideline ARP 4754A (2010) for the development of commercial aircraft with a focus on certification. This document lays out a process for commercial aircraft development and certification which is based largely on SE processes. Comprehensive functional analysis, requirements analysis, and verification are all recommended. Chapter 3 explains that functional analysis can be deleted if the physical architecture is already defined. The lesson is that the existing SE processes should not be ignored or abolished.

The value-added syndrome

The systems engineer will be continually challenged to justify SE on a *value-added* basis. One common justification is that SE will make the design process so smooth that fewer designers will have to fix the mistakes that would have occurred if SE had not been employed. It is always the best policy to justify SE on the basis of value-added.

The importance of content, not form

Another challenge of SE in the commercial world is to keep it lean. As an example, instead of specifications, simple forms or spreadsheets can be substituted. These can be either in hardcopy or electronic form. In any case, in the attempt to keep the process lean, sacrificing content should be avoided.

The terminology conundrum

The one area where time can be needlessly wasted worrying is terminology. For example, well-established SE terminology will have a long-established meaning in the commercial organization. The use of the term *system* to refer to the entire product and not just the electrical system, for example, will be completely unacceptable. The best course of action is to accept the prevailing terminology.

Buy-in from management

SE cannot be incorporated solely from the bottom. Buy-in from top company management is absolutely essential because engineers do not have the authority to change their work priorities or budget or modify their work processes.

Buy-in from management is particularly important with respect to one issue: is SE a technical or a managerial process? The correct answer is that it is both. As explained in Section 13.2 most companies are organized along technical and managerial lines. So how do you implement a process that crosses the technical vs. managerial lines? The recommended approach is to create a SE and integration team (SEIT) that covers both technical and managerial functions. To do this will require strong buy-in from management.

The importance of incremental progress

The most frustrating aspect of being a systems engineer in the commercial environment is the lack of willingness on the part of others to accept all aspects of SE at once. This aspect results from the lack of resources, the *show me* attitude, and the lack of adopted processes. Thus, we are often forced to be satisfied with incremental progress.

The importance of training

There are very few people in the commercial domain who understand the principles of SE. Thus, the establishment of a comprehensive training program is essential to the success of SE.

The importance of striving towards a complete but tailored SE process

This book does not advocate a diluted SE process; however, it does advocate a tailored or adapted process. To be effective, SE should strive to be complete. That is, the functional analysis should attempt to consider *every* function of the product. The requirements analysis should attempt to identify *every* requirement associated with a function. And *every* requirement should be verified. The systems engineer will encounter endless pressure to produce less than is needed. Nevertheless, incremental improvement and thorough training are essential to a complete SE process.

It might be concluded that the adaptation principle described at the beginning of this chapter conflicts with the completeness principle: it does not. The point is that if the deletion of an SE step can be determined to be low risk, then its deletion will not endanger the completeness principle. However, as pointed out above, every deletion should be seriously evaluated.

The adoption of reason

It is difficult to tell a person when their conclusions are not based on reason. Nevertheless, persons in various domains have made conclusions and assumptions that were not based on reason. These conclusions and assumptions, often called paradigms, have sometime led to catastrophic consequences. Jackson (2010, pp. 91–119) provides a set of these paradigms from many sources. One of the most famous of these is quoted by Leveson (1995, p. 57) is called "the Titanic effect." This paradigm states in essence that a system is not necessarily safe just because you believe it is safe. Many disasters including the *Titanic* itself and the *Columbia* and *Challenger* disasters can be traced to the phenomenon of faulty paradigms.

There is no simple solution to resolving how questionable paradigms can be corrected. One cannot simply state to someone, "Think logically" and expect it to happen. The most that can be expected is that within a commercial aircraft organization there should be a system of checks and balances. That is, there should be a system of independent review. Chapter 15 discusses the concept of independent review more thoroughly and asks the question, "What is independence?"

13.2 Adapting the SE Process to the Existing Organization

Before we discuss organizational structures and how SE fits into an organization, we need to stress the point that the organization is a system itself and that the individual departments within a company are the elements of the system. Every department has a product, for example, data that must be transferred to another department so that the receiving department can execute its function, for example, design, test, and so forth.

A discussion of the organizational aspect of SE is rare in the literature with the exception of a discussion of IPTs. Hence, since all aircraft developers have, for the most part, similar organizational structures, and since each element, or department, within an organization has its own unique role in SE, it is hoped that the following discussion will aid the developer in implementing SE within its own organization and within the component parts of that organization.

Hitchins (1993, p. 55) states that an essential property of a system is *cohesion*. Cohesion is the property that all of the parts of a system must interact with all the other parts of the system. In the context of an organization we can say that all the departments must interact with all the other departments. If there is any sort of irregularity among the interactions, then this irregularity will manifest itself in *variability* of the interactions, and the organizational system will drift towards a chaotic pattern in which the final product, the aircraft, is either deficient in quality or experiences higher cost and delayed schedules. Variability can manifest itself in many ways: poor data, lack of data, delayed data, and so forth. Chapter 14 discusses variability in more depth in the context of large-scale system integration (LSSI) and the causes and consequences of it. So making the organization and all its parts work together in harmony is the goal here.

Every commercial aircraft company has an organization. While the names of the departments within these companies may differ from company to company, certain department names will be common across companies. This section will use *typical* department names that should apply to any company. In this section we will capitalize department names as they would be capitalized within companies.

So why are we doing this? The purpose is to show that any existing organizational structure can be adapted to the SE process. The basic principle to be remembered here is that almost all departments have a role in the SE process. This fact may come as a surprise to some who have not yet understood the broader meaning of the word *engineering* as explained in Chapter 1.

The paragraphs below describe what roles specific organizations *should* have with respect to SE. The best of all outcomes is that these organizations *already* perform these roles. If this is the case, then the dictum from Section 13.1 to "respect the SE that already exists" will have been fulfilled. Otherwise, these organizations may need to consider a somewhat expanded role.

Many companies may already have department names that reflect functions that are already recognized as parts of the SE process. This fact supports the rule articulated in Section 13.1, namely the *Existing SE* rule that states that the company should respect and not dilute the SE that already exists. Following are a few typical departments that may already exist within a company:

- Configuration Management. As described in Chapter 12, this department is responsible for controlling both physical and functional configurations.
- Integration. Also, as described in Chapter 12, this department is responsible for assuring that all parts of the aircraft come together in a logical way.

- Integrated Product Development (IPD). Although many standards consider IPD to be a separate and independent process, its basic elements have been part of SE from the beginning. These differences in definitions put aside, the existence of an IPD department within a company is a contribution to SE.

Having said that we do not want to dilute the SE that may exist in these existing organizations, we still have to ask some basic questions: Is the SE in these organizations complete? Do they interact with other organizations in a way consistent with SE principles? We will have to examine these questions as we go along.

Traditional organizations

The challenge of introducing SE into a traditional organization is that most organizations are organized along technical and managerial lines. Jackson (1997) encourages organizations to adopt the SE philosophy and processes while disturbing the current organizational structure as little as possible. Part of the challenge is that, as was stated above, SE includes both technical and associated managerial functions. So it is not an engineering function in the classical sense. There are at least two approaches to solving this dilemma.

The first alternative is not to modify the current organizational structure at all. SE will remain as a function within the Engineering department. This approach will place a large responsibility on SE to educate and collaborate with the managerial departments to explain to them what their roles are. This approach may receive some resistance from these departments and will require strong support from Program Management to make sure these departments understand and execute their responsibilities under this system.

The second alternative is more conventional and in agreement with the broad view of SE in the literature of today. This view would involve the creation of what is generally called the Systems Engineering and Integration Team (SEIT). This organization would include SE and other associated functions, such as Configuration Management and Safety. The important thing about this concept is that it would be at an organizational level above all engineering and management functions and report directly to the program manager. In this way the SEIT would be able to oversee all SE activities, whether technical or managerial.

Either way, almost all departments would have a role in SE one way or the other. Whichever alternative is chosen, though, it must be remembered that the primary consideration is risk. Remember that Figure 13.1 shows that whatever shortcuts are taken, the one with the lowest risk is the right one. Following are a few of the salient organizations and their roles:

Organizations and their SE functions

The task is now to identify typical organizations and ask what SE functions they might perform. So as not to have to repeat this recommendation within the discussion of every organization, it is suggested here that each organization have at least one highly knowledgeable person in SE on their staff. This person would be responsible for advising the organization regarding how to implement SE. Another benefit of this recommendation is to make the minimum impact on the organization and on the organizational structure as a whole.

Program management

SE begins with program management which includes the chief engineer. The primary responsibilities of program management are as follows:

- First, program management must recognize that SE is not simply a subordinate function to program management but that they themselves play an important role in SE.
- Next, as has been stressed before, SE cannot succeed unless all organizations understand their own role in SE. This may be difficult for organizations that are not part of a traditional engineering discipline. It is the responsibility of program management to assure that all organizations understand their role in SE, as outlined in this chapter.
- Among all functions of program management probably the most important and difficult is to provide support for the risk management process as described in Chapter 15. This support consists of the following:
  - defer to expertise in the identification of both technical and non-technical risks and the methods to mitigate them;
  - resist the temptation to invoke managerial prerogative of overriding expert opinions on risk;
  - seek outside independent opinion on risks;
  - take decisive actions to mitigate risks even when costs are involved; these decisions will actually save money in the long run.
- Periodically review requirements metrics, such as technical performance measurements (TPMs), and take action to assure that they are on track.
- As described in Section 4.11, the chief engineer will be one of the key people who decides what requirements are either necessary or unnecessary.

In short, SE will not succeed unless it has total program management support.

Design organizations

Every aircraft company has many design organizations. Typical design organizations are Electrical, Mechanical, Hydraulic, Avionics, and so forth. They have many roles in the SE process, for example:

- They have to write the requirements in the supplier specifications. To do this they must first have a firm grasp on what the requirements are. Some of these requirements will be derived or allocated from higher levels of the aircraft hierarchy as explained in Chapter 4. They will receive assistance from the SE organization regarding the criteria for a valid requirement. The SE organization may need to provide assistance on how to allocate or derive requirements.

- When a modification is being made at the lower level of the aircraft architecture, perhaps level 4 or 5, they will need to verify that the requirements for this modification are in compliance with the top-level requirements, that is, will a modification at level 4 or 5 affect aircraft range or durability, for example? See Section 2.3 for a discussion of the levels of aircraft architecture.
- They will have to assign verification methods to all the requirements as described in Chapter 11 and assign these verification tasks to the appropriate organizations, such as the Test or Quality organizations or the supplier.
- They will have to perform some verification tasks themselves. Primary among these tasks are verification by analysis (including similarity) and verification by inspection. When these tasks are performed, they will have to sign the appropriate documentation.
- When a supplier performs a test or analysis or any other type of verification, they will be responsible for reviewing the supplier's verification results and confirming that they are correct.
- They are responsible for reviewing all verification results and confirming that they comply with the qualification requirements for the aircraft.
- They will have to conduct the required trade studies using the requirements that have been developed.
- Of course, using all of these requirements, they will have to produce a design that meets the requirements.

The Flight Operations organization

It is easy to overlook Flight Operations as a player in SE. Yet this organization employs a product that is essential to safe aircraft operations, the flight manual. The flight manual can be considered an end product to the same extent that the aircraft itself is a product. The flight manual is the product of the same requirements chain that resulted in the aircraft. It is therefore essential that this manual be produced with the same rigorous process that produced the aircraft.

The Marketing organization

Marketing is probably one of the last organizations that would agree that it has a role in SE. However, this is the organization that has direct contact with customers and potential customers. This role of direct contact requires special treatment. Chapter 4 explained the difference between customer needs and product requirements. The Marketing department has the responsibility for capturing customer needs in such a way that they can be transformed into product requirements. For new aircraft this role requires working with the Advanced Design department to achieve this transformation.

The Customer Engineering organization

Aircraft modifications require Customer Engineering to work with Design Engineering to accomplish the goal of modifying the aircraft. Over the lifetime of an aircraft model, the engineering effort of modifying the aircraft may very well exceed the effort to design a new model. As explained in Chapter 4, the important thing to remember is that it is the task of these organizations to determine the true needs of the customer rather than pre-conceived solutions.

The Advanced Design organization

Advanced Design has the primary responsibility for creating designs based on customer needs at the top level. Advanced Design will use methodologies such as the one described in Chapter 8. Advanced Design will determine the architecture of the entire aircraft, such as the number of engines, and so forth. One of the methodologies Advanced Design will use is functional analysis described in Chapter 3. Part of this architecture task is the determination of the number and identities of the subsystems. Advanced Design will be responsible for determining the top-level requirements that will be allocated to the subsystems or derived at the subsystem level. This information will have to be passed to the Design organizations responsible for each of these subsystems.

   Since Advanced Design has the primary responsibility for the architecture of the aircraft, this responsibility will ultimately result in a strong participation of the introduction of *resilience* into the design, as described in Chapter 16. Although resilience is a new aspect of design, many authors, such as Zimmermann et al. (2011, pp. 257–296) have shown that this aspect is of paramount importance in aircraft design. Jackson (2010) has shown that the resilience of a system is primarily accomplished through its architecture.

The Safety organization

More advanced thinkers in safety, such as Leveson (2002) (2006, pp. 95–123) have seen an expanded role of the Safety organization and the safety function itself. Traditional safety standards, for example DoD (2012), focus almost exclusively on the design of the system and its safety. Leveson, for example, places greater emphasis on the organizational aspects of safety. Indeed, the first edition of this book (1997, pp. 126, 139, 153) discusses organizational safety. With respect to the resilience aspect of safety, as discussed in Chapter 16, the emphasis is more than just preventing failures; it is anticipating disruptions and recovering from them.

With this new and expanded view of safety, the role of the Safety organization will remain largely unchanged, that is, to assist the Design Engineering and Advanced Design organizations in implementing this expanded view.

Finally, it is recommended that the Safety organization be an integral part of the Systems Engineering and Integration Team (SEIT) suggested below. This is not to say that Safety is part of SE but to say that the functional binding between the two is so strong that organizing them together would be beneficial to the entire organization. Another strong argument for this grouping is to reflect the high priority have Safety should have in an organization so that instant access to the program manager is possible.

The Certification organization

The Certification organization is a requirements organization of a different sort. This organization deals with regulatory requirements and with the regulatory agency that produces them, such as the FAA or the ICAO. Their job is to interpret these requirements and communicate them to the Design organizations for implementation. This process requires some negotiation with the regulatory agencies who may be, to some extent, flexible in how these requirements can be implemented.

The Systems Engineering and Integration Team (SEIT) organization

The Systems Engineering and Integration Team (SEIT) is the only organization that is unlikely to exist in a present-day commercial aircraft company; although, this concept is so well-known that its existence is not totally out of the question. Secondly, the SEIT may appear to be the only new organization recommended here. This impression is only partially true since parts of it already probably exist within the company. Finally, it is strongly recommended that the SEIT have a direct reporting relationship to the project manager for reasons that will be discussed later. (It is assumed that the company will be divided into multiple projects for each new aircraft being developed.)

So the question to be answered is: Why should these organizations be combined at all? The answer is called *functional binding*. This means that these organizations have such a close interrelationship that keeping them together both functionally and physically enhances their effectiveness. According to Hitchins (1993), entities that are functionally bound have fewer interfaces and are therefore less complex and subject to flawed interactions. So what are the functions that should comprise the SEIT? The following list is only a notional structure which may be modified at the program manager's discretion:

- SE—This is the core organization within the SEIT. It will be discussed more at length below.

- Configuration Management—Most SE texts consider Configuration Management to be part of SE, so it is natural to include this as well.
- Integration—This is also a traditional part of SE.
- Integrated Product Development (IPD)—This function, at least the principal elements of it, are consistent with traditional SE.
- Safety—This function is included primarily because of its importance and its interrelationship with SE.
- Reliability—This function is included because of its aircraft-wide implications and its hierarchical structure.

So the next question to answer is: Why does the SEIT have to report to the project manager? The answer lies in the hierarchical depiction of the aircraft elements as described in Chapter 2 and secondly in the hierarchical nature of the SE approach. SE has often been described as a top-down approach, which is not totally true, but to the extent that it is true, having a bird's eye view of the aircraft systems and their requirements makes SE function as it is supposed to function. The SE organization at the SEIT level assures that the aircraft-level requirements are captured and allocated or derived down to the subordinate subsystems that comprise the aircraft. Chapter 4 describes the process of allocation and derivation of requirements.

The SE organization

The functions of the SE organization are many, but here is a summary of the functions that appear to be most important in the commercial aircraft domain:

First, the SE organization should be recognized as the definitive authority on what SE is and how to implement it. There is a view in some quarters that the "everyone-is-an-SE" is the best approach. This is not totally illogical but this philosophy may lead to an environment in which there are conflicting and even incorrect views of SE. This environment may even lead to conflicting and hence defective design solutions. So, the implementation of this authority function requires that the SE organization write or approach any internal documents pertaining to SE and to provide authoritative training on the subject.

Next, the SE organization will be responsible for providing SE training for key personnel. Among the key personnel are the project manager and the SE focal within each organization listed in this chapter. As mentioned above, part of the adaptation process is to have at least one knowledgeable person in each organization. Training all personnel may prove to be unreasonably expensive. Training of the authors within the design groups for the supplier specifications is especially important.

Training of the project manager and other design group managers requires special attention. These are people who know the SE process and are responsible for making important design decisions. Thorough and intensive training for these leaders is especially important. One technique actually used in practice is to assure that future project and design managers are members of the SE organization. That is to say, they would spend a period of time doing all the SE functions including training themselves. This technique is worthy of consideration.

With regard to requirements the SE organization plays a major role. First, this organization will be the custodian of the requirements tool. A typical tool is DOORS (Dynamic Object-Oriented Requirements System); however, there are others on the market. Chapter 12 discusses the desirable characteristics of such a tool. In the interest of adaptation, the SE organization will not necessarily capture all requirements on all subsystems. However, they will record the ones of interest as the need arises. However, a complete recording of the top-level aircraft requirements, for example, durability and range, is a necessity. As explained in Chapter 4, all other requirements must trace to these requirements. When a requirement becomes official, that is, it is the requirement to which the aircraft or any of its subsystems must be designed to, the Configuration Management group within the SEIT will record and control that requirement. In addition to customer requirements and other requirements derived from them, the SE data base will also include the certification requirements from the regulatory agencies discussed above.

With regard to requirements, a primary responsibility of the SE organization is to review supplier specifications for accuracy and completeness of requirements. If the SE organization does this thoroughly, then the need for requirements reviews will be minimized. The requirements qualities in Section 4.10 will guide the SE organization in their task.

Apart from requirements, the SE organization will need to be available to facilitate trade studies, quality function deployment (QFD), and other SE processes.

Finally, as described in Section 4.11, the systems engineer will be one of the key people who decides what requirements are either necessary or unnecessary.

The Test organization

The Test organization, sometimes called Test and Evaluation, obviously plays an important role in SE since test is one of the four types of requirements verification as explained in Chapter 11. The Test organization may perform tests beyond those demanded by the requirements, but those tests could be considered routine tests that pertain to *routine* or *undocumented* requirements. However, the Test organization does not determine which requirements need to be tested; that is done by the Design organization as described above. The Design organization also determines the pass-fail criteria for the requirements that the Test organization must use. Sometimes tests are witnessed by the Quality organization to determine whether the test passed or failed.

Of course, there are test organizations within the supplier companies as well. The function is basically the same, but it is important that the tests be witnessed by Quality personnel from the developer organization.

Reliability organization

Reliability is a specialty engineering subject of long standing. However, within an SE context it has a special role; that is why it is part of the SEIT. It is not simply the role of the reliability engineers to determine what the reliability of the aircraft it; it is the responsibility this group to specify what the reliability of each element of the aircraft should be. To do this the reliability engineers need to allocate the reliability from the aircraft level to each of the lower levels of the aircraft hierarchy. This is not to say that they don't do this already; if they do, then this is another example of the "SE that exists" and does not need special attention.

Specialty Engineering organizations

Within a commercial aircraft company there are many specialty engineering organizations. These include human factors, maintainability, lightning, electromagnetic interference, and others. The important principle to keep in mind here is that these organizations are not independent entities. They are tied together by the concept that the aircraft is a system and that the engineering disciplines that make the parts work together are a system as well and the *cohesion* between these disciplines needs to be maintained as well.

All of the specialty engineering organizations together comprise a group which is often called an AIT (analysis and integration team). An AIT is similar to an IPT except that it does not represent a particular element or subsystem of an aircraft as described in Chapter 12. A specialty organization may often consist of no more than one or two engineers, so it is impractical assign one specialty engineer to each IPT. They have to divide their time among multiple IPTs.

The main interface for the Specialty engineering organizations is the Design organizations who are responsible for implementing and verifying the specialty engineering requirements.

As described in Section 4.11, the specialty engineer will be one of the key people who decides what requirements are either necessary or unnecessary.

Supplier Management organization

Supplier Management, like Marketing, does not think of itself as an engineering organization; and it is not. Yet it plays an important role in SE. Chapter 14 discusses the importance of reducing the *variability* among the elements of the supply chain, where variability is the deviation in the quality of the contracts, specifications, and the resulting supplier products. Supplier Management is sometimes called Procurement.

Supplier Management does not write contracts nor does it write specifications. The Contracts and the Design organizations do those things: Supplier Management's role is to make sure they get done and done correctly. Supplier Management has other key functions, such as to assure that potential suppliers are financially sound and have a history of producing quality products.

The Contracts organization

In practice there may be two Contracts organizations, one for contracts with the airline customers and one for contracts with the suppliers. It doesn't matter. The important point is that a contract is one of the mechanisms that reduces the variability between the elements of the large-scale system (LSS) which includes the developer, the customer, and the suppliers.

Contracts with the airline customer, of course, guarantee what the customer will get, how far it will fly, how much it will weigh, and so forth. If the aircraft does not meet these requirements, there will be financial penalties on the aircraft developer. Overweight aircraft is one of the more frequent examples of such deficiencies. In accordance with the equations for aircraft performance, for each pound of overweight aircraft, there will be a corresponding deficiency in range and payload capability. The important point, though, is that these same requirements will be the same as the top-level requirements tracked by the SE organization in their requirement tool as described above.

Of equal importance are contracts with the suppliers. As discussed in Chapter 14, this is one of the primary mechanisms for reducing the variability in the large-scale supply chain system. As stated in Chapter 14, the important principle is to keep contractual requirements and technical requirements separate. In practice, this is difficult to do; nevertheless, there is risk in combining the two.

Also discussed in Chapter 14 is the importance of internal contracts, that is, contracts between different divisions of the aircraft company. Whether these internal contracts are written by the same Contracts organization may vary from company to company, but the important principle is that these contracts should be as rigorously written as external contracts.

The Production organization

Another organization unlikely to associate itself with SE is Production. Yet Production is an essential phase of the creation of a commercial aircraft, and the physical system actually produced should match the requirements in the SE requirements data base and the system actually designed by the Design organization.

The observer can appreciate the risks of not producing the product actually desired by looking at the many steps the requirements have to traverse to result in a physical product. We have already seen how customer needs are converted into product requirement and how the Design organization creates a design that meets these requirements. To become a product the requirements have to go through two more stages: First, they go to an organization called Planning who create the instructions for how to assemble the final product. The production mechanics then produce the aircraft using these instructions. The potential for mistakes increases as the number of steps increases. The insurance against mistakes is called verification as described in Chapter 11. At the Production level this verification is performed by the Quality organization.

As a simple example, the mechanics are required to install the data lines and electrical conduits at a prescribed spacing due to EMI requirements. If, due to insufficient training or any other reason, the mechanic fails to achieve this spacing, the requirements will not be met and equipment failures may occur. Once again, the Quality organization is responsible for assuring that the requirements are met.

One essential last duty of the Production organization is to communicate with the Design organization when the Production organization perceives that the design is not producible. This may be a simple issue, such as a door that will not close. In any event, this issue needs to be reported immediately to correct the situation and prevent a repetition of the problem.

The Planning organization

As described the Planning organization is responsible for passing on instructions to the Production organization that reflect the requirements at the product level. Human error can occur at this level also. However, the Quality organization once again has the responsibility for assuring that the requirements are met.

The Quality organization

The Quality organization has broad responsibilities. However, these responsibilities all fall under the umbrella category of verification, in particular, verification by inspection as discussed in Chapter 11.

They can, for example, inspect drawings to assure that they meet the requirements of the specifications. They can witness qualification tests at supplier sites to assure that the supplier product meets the requirements. They can inspect the Planning documents to assure that they meet the design requirements. Finally, they can inspect the final aircraft product to assure all Quality organizations in all companies may not do the tasks listed above. These are only typical tasks that assure that the final product meets the customer requirements.

The Maintenance organization

Of course the Maintenance organization is not part of the commercial aircraft development company but rather part of the airline company or a supplier to the airline company. The maintenance manual produced by the aircraft company can be seen as a requirements document just as much as a procurement specification. The airline company is contractually obligated to conduct its maintenance operations in accordance with this manual.

The criticality of maintenance to the aircraft enterprise cannot be over emphasized. Reason (1997, p. 88) cites defective maintenance as one of the primary causes of aircraft accidents. He lists the 1979 American Flight 191 as one of many examples.

The practice of outsourcing maintenance to outside organizations has contributed an extra layer of risk to aircraft operations. Chapter 14 shows how the increased number of organizational interfaces adds risks due to the increased variability of communications between organizations.

Summary of organizational responsibilities

The astute reader will by now have detected that there is a central theme in the above description of organizational responsibilities. The theme is that all commercial aircraft organizations play a role in converting customer requirements into a physical aircraft product and verifying that the aircraft meets these requirements. The goal of this section is to show that even existing organizations that may not have totally understood this theme can fit into the pattern and accomplish this objective.

14

Large-Scale System Integration

The toe bone's connected to the foot bone,
The foot bone's connected to the ankle bone,
The ankle bone's connected to the leg bone,
Now shake dem skeleton bones!

"Dem Bones," American spiritual

There is no topic more important in modern commercial aircraft development than large-scale system integration (LSSI). The large-scale system (LSS) in this instance is the global supply chain. Every aircraft developer contracts with suppliers all over the world for components and other aircraft parts. This international procurement is commonly known as *outsourcing*, which will be discussed in this chapter. Commercial aircraft are part of other LSS, such as the global navigation system and the air traffic control (ATC) system. But this chapter will focus primarily on the global supply chain since it is the sources of the greatest risks and is in need of the greatest degree of management, both technically and organizationally.

14.1 The System of Systems View

Before we can address the integration of the global supply chain system, it is necessary to view it as a single system with interconnections and not a set of disconnected pieces and that the developer has the responsibility for assuring this integration. In SE this system is also known as a system of systems (SoS). This is because the suppliers themselves are systems that are independently developed. When a set of independently developed systems are required to function together to achieve a common goal, the resulting set of systems is called a SoS. More on systems of systems can be found in Jamshidi (2009).

14.2 Outsourcing

Outsourcing is the procurement of parts and services through organizations outside the developer's own organization. Very often this procurement of parts and services is through companies in other countries. There are four primary reasons for outsourcing:

- Suppliers can provide products or services at a lower cost. This is often the case for suppliers in countries where labor rates are lower. This type of outsourcing does not only apply to product development but also to maintenance. When maintenance is performed in low-cost distant countries, this practice increases the chance of poorer quality work and also makes it more difficult for regulatory agencies to inspect and approve the maintenance facilities and procedures. This type of outsourcing is performed by airline companies rather than the developer thus adding layers to the procurement process. These additional layers also add risk as will be shown later in this chapter.

- Suppliers can provide a unique or perhaps technically superior product. There are very few engine suppliers in the world. So developers have to rely on one of these for their engines regardless of where they are located.
- Suppliers may be located in a country who is a potential buyer for the aircraft. So the developer sells the aircraft to a country in exchange for the purchase of parts or services. The purchased products do not even have to be aircraft parts. They can be soccer balls or cans of ham. This practice is called *off-set* marketing. This practice sometimes occurs because the country of purchase may have a weak currency and may prefer to buy the planes with goods or services rather than cash. The supplier country also seeks to create jobs in their own country; this factor provides the incentive to them to negotiate an outsourcing contract.
- The developer may lack the investment capital to finance a totally internally developed product; hence they depend on other companies, generally called partners, to help finance the aircraft and hence share the cost risk.

Other possible reasons for outsourcing may include:

- Investors may *perceive* that outsourcing will reduce the risks of development. The discussion in Section 14.2 will show that the increase in the number of organizational interfaces will actually increase the risk.
- Sometimes tax incentives in foreign countries may have the effect of subsidizing development in those countries.
- Reducing internal labor costs also reduces the amount of other long-term costs like benefits, such as pensions.

Some of these reasons may be in conflict with each other. For example, if a product is outsourced to another country for sales purposes, that country may actually have labor rates greater than the developer's country.

This is not to say that outsourcing is necessarily bad. All commercial aircraft companies do it for one or more of the reasons cited above. The point is that outsourcing should only be done moderately. Excessive outsourcing can have serious consequences, that is, it will increase risks rather than decrease as was the expectation. The reasons will be delineated below.

From an organizational perspective the degree of outsourcing will be the responsibility of executive management. It is also their responsibility to remain fully informed about the risks of excessive outsourcing. It will be the responsibility of the supplier management organization to determine the soundness of the supplier organization. Engineering will determine the acceptability of the supplier's product. Sillitto (2010) emphasizes the need for continuous feedback in the system. Most importantly there needs to be feedback between these tiers. Additionally there needs to be feedback between the suppliers and the developer.

14.3 Complexity and How It Increases Risks

To understand the risks of outsourcing and LSSI it is necessary to understand the concept of complexity at least some of its primary tenets. There have been many books written on complexity, for example, Page (2011). We will only discuss a few of the important aspects of complexity to show how complexity increases risk, how it relates to LSSI, and how the commercial aircraft developer can decrease complexity and decrease risk.

The Large-Scale System (LSS) and complexity

Sillitto (2010) states that one of the primary attributes of a LSS is complexity. He says that these systems are "dominated by emergence and exist in a state of constant reconfiguration and evolution." Sillitto calls these LSSs *wicked* systems. For the uninitiated, emergence, according to Checkland (1999), is any property of a system that cannot be attributed to a single element, such as a single supplier. Emergence can be the result of a failure, schedule slip, or cost overrun. Emergence can also be beneficial, but the focus in this chapter will on detrimental emergence. It takes the whole LSS to create emergence. Emergence is a basic feature of complex systems. Commercial aircraft supply chain systems qualify as complex systems. So the basic question is: How do we manage complexity and reduce the risks associated with it?

Some features of complexity

Of course, the more system elements, suppliers, there are the more complex the system will be. So the obvious way to reduce complexity is to reduce the number of suppliers. However, there can be limitations to this process; we have already noted that outsourcing is a necessary part of aircraft development. So there is a limit to how much the developer can reduce complexity.

Humans in the system

It is axiomatic of SE that humans are not simply operators of the system or designers of the system; they are elements of the system itself. This is especially true of the supply chain system. There are many humans in the supply chain system. Humans alone do not make the system complex. However, there is general agreement that systems with many human elements are complex and suffer from the variability of human performance. This is not to say that humans should be replaced with technical components or that humans are necessarily untrustworthy. It is only to say that any interaction between humans can be variable, and it is this variability that needs to be managed as discussed below in the section called Variability between Elements. It is this variability that is the vulnerability of the entire global supply chain system.

According to Sillitto, human systems are regarded as *soft systems*. The products produced by suppliers are called *hard systems*. Sillitto says that hard systems exist within soft systems. This multiplicity of system types is the factor that makes the supply chain complex. Regardless of the tiers, both systems strive for a common purpose, that is, a successful commercial aircraft system. This success can only be achieved if the variability is reduced or eliminated between the systems.

Organizational interfaces

There are two more aspects of complexity that are often mentioned in the literature, for example by Marczyk (2009). One is the number of interfaces, and the other is the variability among the elements. Both of these are important to reducing the risks associated with complexity. We will discuss interfaces first. Chapter 6 has already discussed interfaces, but the emphasis there was interfaces internal to the aircraft. The interfaces here are generally organizational interfaces, that is, interfaces among the elements of the supply chain including suppliers, maintainers, and the developer.

Jackson (2010) discusses how the numerous interfaces among organizations have led to communication breakdown that was responsible for some of the major air disasters. For example, first there is the interface between the developer and the airline. Secondly, there is the interface between the airline and the maintenance organization. It does not matter whether the maintenance organization is part of the airline or a separate supplier. Either way, there is an organizational interface. For example, the American Flight 193 accident was caused primarily by improper engine installations. This was an internal organizational interface. The ValuJet accident was caused by the actions of an external maintenance supplier. In short, multiple organizational interfaces can add risk to the aircraft operation. Reason (1997) notes that insufficient maintenance is the root cause of a large percentage of accidents. This does not imply that the developer, the airline, or the maintenance organizations are negligent. It only implies that the increased number of organizational interfaces increases the risk. Figure 14.1 shows schematically how the maintenance organization is separated from the developer by multiple organizational interfaces.

This diagram shows how the regulatory agency, the Federal Aviation Authority (FAA) in the United States, adds at least three more organizational interfaces to the system. These agencies have an active and essential role in regulating the products of the developer, the airline, and the maintenance organization.

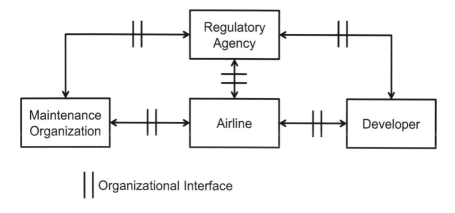

Organizational Interface

Figure 14.1 External organizational interfaces

Internal organizational interfaces

Internal to a developer's organization there are perhaps hundreds of organizational interfaces between, for example, design, test, supplier management, and so forth. Each of these interfaces is a potential source of information errors. Figure 14.2 depicts a typical set of organizational interfaces over which errors frequently occur. This set involves three functions:

- Engineering. This is where a product or a subsystem of the aircraft is designed.
- Planning. The planning function takes the product as designed in engineering and plans the steps for producing the product. This step can involve steps such as pick up drill and drill hole.
- Product. This function executes the instructions of the planner.

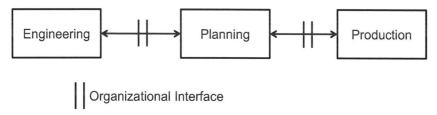

Organizational Interface

Figure 14.2 Typical internal organizational interfaces

Variability between or among elements

An important contributor to complexity is variability between or among elements as described by Marczyk (2009). An obvious question to ask is: Variability in what? A general term is variability in connections. Connections can consist of any sort of communications, either verbal or written. Variability in specified parameters, text results, contractual provisions, and so forth also count. Variability between elements can take many forms. For example, an ambiguous performance specification or contract between the developer and a supplier can constitute variability. This one aspect alone argues for intensely rigorous specifications and contracts. These will be discussed later in this chapter. Variability can consist of loose oversight of maintenance procedures. Variability can either consist of inaccurate information or an absence of information.

A potential source of variability is the differences in cultures between the developer and various suppliers. This variability can result from differences in languages and also in business practices and customs. For example, Applegate (1998) discusses the development of an aircraft in Indonesia. Factors mentioned by Applegate include the many languages spoken in that country.

The Haddon-Cave report (2009) on the Nimrod crash also discusses risks due to overly complex supply chains.

Multiple regulatory agencies

The existence of multiple regulatory agencies adds complexity and variability to the connections of the global supply chain system. Sometimes the regulations among these agencies can be in conflict. The developer is then responsible for resolving these conflicts. Of course, the multiple regulatory agencies will exist whether the developer practices outsourcing or not. However, when these multiple regulations are added on top of an already complex organizational structure, the complexity is then compounded making the risks of outsourcing even greater.

Intermediate stability

Rechtin (1991) emphasizes the importance of *intermediate stability*, that is, the reduction of variability, not only in the whole system, but the pieces of it as well. In this chapter the whole system is the supply chain system. In simple terms this can mean that stability should be strived for between the individual suppliers and the developer. So when managing variability as described in Section 14.3, stability between individual system elements should be kept in mind.

Importance of supplier management

Rigorous supplier management is the key factor in achieving stability in the supply chain system. Within this management, rigorous contracts and specifications and their implementation and enforcement are the important aspects. This section elaborates on those aspects.

Consistent and thorough management of suppliers is central to the reduction in variability. Section 12.10 of the main body of this book discusses the function of supplier management. Supplier management is especially important in LSSI. Traditionally supplier management may have been thought of as a *managerial* function rather than a *technical* function. Nevertheless, supplier management plays an important role in SE and especially in LSSI.

Jackson (2010) provides a list of *paradigms* that are common in industry, not just commercial aircraft. Paradigms are, in this case, ways of thinking or mindsets that are risk inducing. That is, the mindset itself may lead to negative consequences if implemented. One of these is the *independent supplier* paradigm. This is the belief that suppliers can figure out what to build without any requirements from the developer. It is safe to say that such a business practice is not consistent with the rigor that LSSI would demand.

The importance of language

The importance of language between developer and suppliers cannot be underestimated. Today commercial aircraft are being built in different countries with different languages. In addition, suppliers are located all over the world. Total fluency in a common language is essential in contract negotiations and preparations and in specifications. Practitioners in these areas need to have total fluency in a common language. Although the common language is very often English, this is not always the case. Verbal competency is not sufficient; total fluency is required to detect the slightest nuances in meaning of any given thought. If the developer or the supplier needs to hire qualified interpreters or translators, this should be done.

Internal suppliers

Often large companies have many divisions located in different cities and perhaps different countries. When one of these divisions makes a product or component of the aircraft, these divisions are then assuming the role of supplier. These divisions are then internal suppliers rather than external suppliers. When products are procured internally, there has to be some internal documentation to procure these products. It is possible this documentation does not even pass through the supplier management department or the contracts department.

Here is the source of another risk. If the internal documentation does not meet the standards of external specifications and contracts, there is the risk that the product itself will not meet the same standards that an external product would meet. As a general principle this documentation should meet those standards.

14.4 Managing the Risks of a Large-Scale System (LSS)

The discussions above have identified two principal root causes of risks for LSSs. These risks are associated with the global supply chain system typical of the commercial aircraft industry. The two sources are as follows:

- the multiplicity of organizational interfaces;

- the variability of connections between the elements of the supply chain system.

Managing organizational interfaces

The above discussions point out the risks inherent in an excessive number of organizational interfaces. There are only a limited number of ways to manage this number.

The first and obvious way to reduce the number of ways is by reducing the number of suppliers, that is reducing the extent of outsourcing. As was pointed out above, there are good reasons for outsourcing. So the main goal is to minimize the extent of outsourcing. There are only two ways to do this:

1. Perform component development internally. This can only be done if the developer has the capability.
2. Utilize internal suppliers. Similarly, this option has its risks as well. Internal suppliers can only be used if the internal specifications and cross-organizational agreements have the same rigor as external specifications and contracts.
3. Engaging one supplier to integrate an entire subsystem, for example, the avionics system or the propulsion system. This way the single supplier can assure the minimum variability among the components in that subsystem. This responsibility would have to be spelled out contractually.
4. Using multiple-source suppliers for critical equipment. A good practice is to have at least two suppliers, and even more whenever possible, in order to be able to challenge them in term of cost at least and never face a situation of a delivery interruption.

After these options have been exhausted, the developer must concentrate on reducing the variability among the elements of the global supply chain elements as described below.

Managing variability of connections

Reduction of variability through supplier management involves two principal thrusts: contractual management and management of specifications. Section 12.9 of this book discusses the role of supplier management. The main point of that section was that the supplier management process should be rigorous, both technically and managerially. Following are two rules for reducing variability of technical and contractual documents for suppliers.

Rule 1: Supplier contracts should contain only contractual information and should not contain any technical requirements
The reason for this rule is that contracts are priced on the work done, not on the performance of the product. Furthermore, if a technical requirement appears in

the contract, there would be no link to the verification matrix described in Chapter 11. So the verification method would not be clear. In fact, it may not even exist. The contract must have a clear reference to the specification that will provide the information about how the product should perform.

Rule 2: Supplier specifications should contain only technical requirements and should not contain any contractual language

The reason for this rule is that if a work statement appears in the specification, there would be no way to price it since all pricing statements are in the contract.

In short, adherence to these two rules will add rigor to the reduction in variability of all the aspects of supplier procurement and hence reduce the complexity of the global supply chain system and the risk of significant consequences, such as technical, schedule, and cost.

Reducing the variability in supplier contracts

Continuing with Rule 1, here are a few topics that are essential to supplier contracts. These topics should apply to both external and internal contracts:

1. The contract should state what work is to be done by the supplier and over what time and how the product should be delivered to the developer.
2. The contract should list any other milestones or events in which the supplier is involved. For design reviews, the contract should state who will conduct the reviews and who will approve the outcomes, normally the developer and not the supplier.
3. The contract should state who will conduct product verification including qualification tests and who will approve the outcomes. This topic includes the five classic verification methods also described in Chapter 11:
   • Analysis—If the supplier has to verify a requirement by analysis, this should be stated in the contract. If the developer has to perform this analysis, the contract should state this fact also. Verification by analysis includes verification by similarity, a common method of verification. Other common analyses include structural analysis and aerodynamic analysis.

   There is an inherent risk in analysis by similarity. This method is only valid if the product is compared to an equivalent product in the same environment as the product being analyzed. Among verification methods similarity is the one that is least understood and most prone to a lack of rigor.
   • Inspection—Either the supplier or the developer can perform verification by inspection. The developer is responsible for final approval.
   • Demonstration—Either the supplier or developer can perform product demonstrations.

- Tests—Normally the supplier would conduct product tests before delivery to the developer. In any case the developer must approve the final result.
- Conformity—Conformity is the assurance that all aircraft parts are in compliance with drawings. The conformity process can be time consuming, but it is an essential process.

The verification method for each requirement will appear in the supplier specification. In accordance with SE principles, all requirements must be verified by at least one of the methods.

Reducing variability in specification quality

Here are a few things to keep in mind when developing supplier specifications:

1. Rigorous requirements are essential to a quality specification. Books have been written on the subject of how to write requirements. Hooks and Farry (2001) is one example of how to write good requirements. Requirements written by untrained persons tend to be ambiguous.
2. Don't forget traceability. No requirement exists in isolation. Every requirement in a supplier specification either flows down or is derived from aircraft-level requirements. See Chapter 2 for what is meant by levels.
3. Make sure your requirements are complete. Don't leave any blanks in the specification. Environmental requirements are especially vulnerable to omission. The environment inside the engine nacelle is very different from the environment inside the cargo bay.
4. Make sure your specification has a good verification matrix. See Chapter 11 of this book for a discussion of verification. From an LSSI perspective, it is important to specify what verification methods the supplier will employ and who will approve the results.

Managing multi-tier suppliers

Suppliers sometimes come in multi-tiers. That is to say, a tier 1 supplier is one that reports directly to the developer. A tier 2 supplier is one that reports to the tier 1 supplier. For example, the aircraft engine manufacturer is a tier 1 supplier because the engine fits directly into the aircraft. In the engine pod there may be a valve attached to the engine. The manufacturer of the valve is a tier 2 supplier. The situation of multi-tier suppliers can lead to risks if the requirements flow down is not rigorous. Figure 14.3 is a schematic that shows the relation among the developer and the tier 1 supplier and the tier 2 supplier.

This schematic shows that the relation between the developer and the first tier supplier is exactly the same as between the first tier supplier and the second

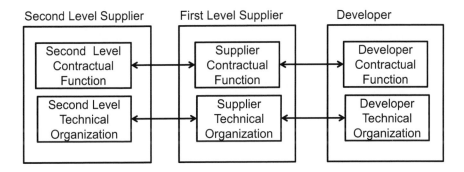

Figure 14.3 Multi-tier supplier relationships

tier supplier. There may be third tier suppliers and fourth tier suppliers but the principle is the same.

This schematic also shows the dual relationship between the tiers, the contractual tier and the specification tier as explained in the paragraphs above. So what are the expectations about multi-tier supplier?

1. The developer can expect that the relationship between the first and second tier suppliers will be exactly the same as between the developer and the first tier supplier.
2. The developer can expect to verify that the second tier contracts and specifications meet the rigorous requirements described above.
3. The developer can expect that the third and fourth tier contracts and specifications will also meet these requirements.

As a simple example described above, the environment for the valve inside the engine pod will be very different from the environment of the engine or any other part of the aircraft. The developer has the responsibility for assuring that requirements such as these are flowed down to all tiers of suppliers.

An accepted way to assure that these multi-level supplier problems are avoided is to perform whole aircraft integration tests before the final aircraft is delivered. This method would, of course, be more expensive, but the risks would be considerably reduced.

Supplier responsibilities

It should not be assumed that the responsibility for quality specifications belongs only to the developer. The suppliers themselves need to provide feedback to the developer regarding:

- missing requirements;
- unclear requirements;

- requirements that are not achievable;
- unnecessary requirements.

The developer needs to provide sufficient slack in the procurement schedule to allow the supplier to provide this information.

14.5   Other Large-Scale System (LSS) Principles to Apply to Commercial Aircraft

Sillitto (2010) recommends that we pay attention to the stability of the LSS and apply measures of stability to these systems. By stability Sillitto means what we have called the lack of variability. Anyone who has worked in the commercial aircraft world knows how all aspects of the system are in a state of constant flux. There are new customers with new requirements and new suppliers all the time. More importantly applicable to this chapter there may be new organizational interfaces and variability at all tiers. There may be variability in contract provisions and contracts. The introduction of new technologies, which happens on a continuing basis, adds new levels of variability and hence instability. Recent years have seen the introduction of new technologies, such as composite structures, electronic generators, and envelope control.

So how can stability be measured? On the simplest level, there are qualitative evaluations. At the macro level the developer can look at the number of outsourced suppliers to see if they exceed the normal number. Then they can look into specifications and contracts to see what changes are occurring.

On a more sophisticated level Marczyk (2009) has developed an algorithm to measure complexity. He can apply this algorithm to the entire global supply chain system and determine how close to the *tipping point* it is, that is how close to the point of disintegration.

Sillitto also recommends that we pay close attention to the node and web architecture. This architecture is a good depiction of organizational interfaces. The fewer organizational interfaces the more stable the system. There is another method for analyzing the node and web architecture and for minimizing complexity of systems that includes both humans and technological elements. This is the optimized N^2 diagram described by Hitchins (1993). In this method the analyst arranges all elements of the system on an N^2 diagram a depiction well-known to systems engineers. The method allows the analyst to rearrange the elements until all the relationships are minimized. This method results in a minimally complex system.

14.6   Summary

We have described the global supply chain system as a LSS in a SE context. This system is characterized as a complex system in which instability and resultant cost, schedule, and technical risks are incurred as a result of two major factors: an excessive number of organizational interfaces and the variability between elements of the system. In this case the elements are the developer, the suppliers, and the regulatory agency. It is shown that this variability can be minimized through rigorous control of suppliers both managerially and technically. Stability among individual elements is also emphasized. It is concluded that adherence to the rules outlined in this chapter will assist in reducing the degree of outsourcing and the associated risks.

15

Risk Management

As a broad definition of risk, risk is anything undesirable that may happen as the result of either an internal or external cause. This is the consequence that occurs *if nothing is done to prevent it from happening.*

Risk management is one of the most difficult tasks to accomplish especially in a commercial aircraft environment where the demands of the marketplace and schedule constraints often place it low on the priority list. Yet its importance cannot be underestimated. This chapter will, first, provide some of the essential principles of risk management and the steps for accomplishing it. In addition, it will provide some hypothetical examples of actual risks and their possible consequences, their warning signs, and ways to mitigate the consequences. For a more comprehensive description of the risk management process, the reader can refer to, for example, Conrow (2003). However, this chapter will provide some guidance on how this process can be adapted to the commercial aircraft domain.

Not to be overlooked are the cultural obstacles to risk management. Vaughn (1997, pp. 96–111) provides the most compelling summary of this issue relative to the *Challenger* disaster in which she coined the phrase "normalization of deviance." In addition, the Columbia Accident Investigation Report (NASA 2003, p. 189) described the "broken safety culture" as a principal contributor to that accident. Finally, Conrow (2003, pp. 122–124) provides a comprehensive list of "negative attitudes" that exacerbate the risks that are already there. While none of these sources provides a clear solution to the cultural aspects, the fact remains that they remain an impediment to the effective operation of a commercial aircraft organization and should be treated seriously.

ARP5754A (2010) discusses risk with a focus on certification and safety. That is, the risks discussed in that document are almost exclusively technical. The scope of this chapter is broader. This chapter discusses risk related to the total development of the aircraft including technical, cost, and schedule risks and the interrelationships among those risks.

15.1 Overview of Risk Management

There are three broad categories of risk: cost, schedule, and technical. Because of its certification implications, safety risk as discussed in Section 10.2 is treated as a separate category in the aircraft industry. The demands of the commercial aircraft industry are that all risk needs to be addressed and treated. A single risk may have all three types of consequences, or perhaps only one, or perhaps two. Hence, it is not appropriate to attempt to assign responsibilities for risk management to technical and managerial organizations. Responsibility should reside in a single technical–managerial organization, such as SE.

Risk management can minimize or eliminate many risks. It can highlight areas of uncertainty or false confidence. It provides a means of deciding the best course of action. It can provide a means of early warning of problems. It can provide a means for plan changes. Finally, and most importantly, it can increase management and airlines' customer confidence.

There are many ways of analyzing and managing risk; for example, the *NASA Systems Engineering Handbook* (1995) describes probabilistic risk assessment (PRA). The following simplified risk management process is both a manageable and effective process for commercial application:

Identify risks

The first step is to develop and document a risk statement. The risk statement should include the source of the uncertainty and the potential consequence. Each risk should be identified as one of the three categories above:

1. *Technical risk*   The uncertainty of achieving program requirements of function, performance, and operability within the planned cost and schedule.
2. *Schedule risk*   The uncertainty in achieving the program schedule if none of the technical risks should materialize.
3. *Cost risk*   The uncertainty in achieving the cost budgets if none of the technical and none of the schedule risks should materialize.

As a general rule, the best method of identifying risks is by using checklists and templates. In interviews, such phrases of uncertainty, such as "not sure," or "don't know" are sure clues.

Analyze risks

The grid of Figure 15.1 has been found useful in risk analysis. A common practice is to separate the risks into three categories: low, medium, and high. The general practice is to manage only those risks that are medium or high. One might ask why the two corners of the grid do not have the same level of risk. As this chart shows, one corner is low risk and the other is high risk. A logical conclusion is that high-consequence, low-likelihood risks are considered more serious than low-consequence, high-likelihood risks. This conclusion is in agreement with the fact that many disasters of high consequence are the result of low-probability events.

A key aspect of risk analysis is that the risks should be assigned and validated by persons other than the component designers or owners. An effective approach is to use an integrated product team (IPT) consisting of the designers and non-advocates, that is, impartial observers.

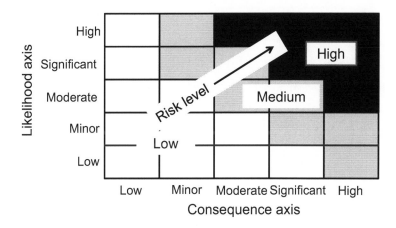

Figure 15.1   The risk grid

To determine where any given component fits into the grid of Figure 15.1 the templates of Tables 15.1 and 15.2 are useful. Every attempt should be made to base the risks on as quantitative a basis as possible. For example, if a supplier has a track record of delivering a product on time only 50 percent of the time, then the likelihood should be judged as moderate.

Table 15.1   Likelihood template

Likelihood Level	Commonly Applied Probabilities	Description
Low	<1%	Desired results very unlikely to happen
Minor	<10%	Desired results have low likelihood
Moderate	~50%	Desired results likely to happen
Significant	>90%	Desired results highly likely to happen
High	>99%	Desired results near certainty of occurrence

Table 15.2   Consequence template

Consequence Level	Description
Low	Little or no impact
Minor	Problem that could easily be handled with available resources
Moderate	Problem that requires handling with additional resources
Significant	Significant change to program viability if not addressed
High	Problem that could result in program termination if not addressed

Develop risk handling plans

If steps are taken in advance, the likelihood or the level of the consequence happening or will be reduced. These steps are normally called *handling* methods.
   The commonly recognized risk handling methods are as follows:

- Mitigation. A risk can be mitigated either by reducing the level of the possible consequence or by reducing its probability of occurrence.
- Acceptance. The developer can accept the risk without any action provided the probability or consequence is extremely low.
- Transfer. The risk can be transferred to another body, such as the customer, the supplier, or the regulatory agency, providing this can be done within the bounds of legal and contractual constraints.

However, as an example of risk transfer, if a customer demands a specific piece of commercial off-the-shelf (COTS) equipment and does not provide evidence that the equipment in flight qualified, then the customer is accepting a *de facto* transfer of the risk. That is to say, if the equipment should fail, the customer is accepting the responsibility for any consequences that may result.

- Analysis. Strictly speaking one cannot handle risk by analysis. However, in the real world facts emerge that enable the analyst to reassess the risk from time to time. Of course, analysis is important as time goes on. Risk must be reassessed after each mitigation step is performed.

Track risks

Risk tracking can be accomplished by common unsophisticated computer applications. Risk tracking provides the data needed to make the decisions outlined in the risk reduction plan. A key aspect of risk tracking is to track only the moderate and high risks. This technique will keep the task manageable and effective.

Holistic Risk Management

Advanced risk thinkers have begun to understand that risk management is complex, that is, the many factors that cause risk may be interrelated. For readers interested in these advanced aspects, the book *Holistic Risk Management in Practice* by Hopkin (2002) is recommended. Among other aspects, holistic risk management integrates aspects of risk across multiple business units. As shown in Chapter 14, since the commercial aircraft system can be seen as a large-scale system (LSS) including the developer and many suppliers, this aspect is particularly relevant to the commercial aircraft domain. Although this book was intended for the financial industry, its principles still apply to commercial aircraft.

15.2   Types of Consequences

As Section 15.1 explained, there are three basic types of consequences:

- A typical cost consequence is when the entire aircraft is delivered to the airline customer late, that is, later than the agree-to contract called for.
- A typical cost consequence is when the cost to build the aircraft is greater than anticipated.
- A typical technical consequence is when the aircraft does not deliver the performance that it was designed to deliver. For example, the range may not be as great as expected, or the weight is greater than expected.

These consequences may occur either at the aircraft level, as described in the above examples, or they may occur at the supplier level. For example, the supplier may deliver their product late, or it may fail the qualification test, a technical consequence.

In addition, these consequences may occur singly, or they may occur in combination. For example, if the aircraft is delivered late, there may be a cost penalty in the contract that the developer has to pay to the customer. If the aircraft does not meet its contractual performance, there may be a schedule penalty to solve the problem and a cost consequence to pay for solving the problem.

15.3   Root Causes of Risks

Risks may be external, that is, out of the control of the developer. Or they may be internal, that is, the result of decisions made within the developer's organization. It is impossible to list all of the risks that a commercial aircraft enterprise may encounter, but the following are typical.

External causes

- The basic top-level external risk is market risk. Does the customer want more range, more seats, or speed? Or has the market dried up altogether? The inability to predict these market factors can be critical.
- There are unexpected technological risks. In spite of the latest ability to establish durability requirements, cracks due to metal fatigue or composite material delamination may start appearing the fuselage. Once built these types of flaws are difficult to fix. Cost, schedule, and performance will suffer.
- Supplier risks are the most difficult. Suppliers with a spotless reputation may deliver a product that is lacking in performance or late in delivery.

Internal causes

Internally caused risks are generally the result of optimistic decisions about anything, for example, the time to develop and the capability of suppliers.

- Excessive outsourcing. Chapter 14 is devoted to the risks associated with excessive outsourcing and failure to manage the variability in contracts and specifications.
- Schedule optimism. Risks occur as a result of optimism pertaining to the development of, for example, internal components, such as electronic modules. A model that is estimated to take, for example, two years may wind up taking three years. The primary consequences are both cost and schedule.
- Lack of technical expertise. The developer may embark on a new technology, such as composite structures, without a complete understanding of some of the issues associated with it, such as delamination.
- Supplier evaluation. Risks may occur as a result of the inadequate evaluation of a supplier. A supplier may be contracted to develop a new technology that has never been built before. All three: cost, schedule, and performance risks will result.
- Flawed assumptions. Risks can occur when suppliers are given obsolete specifications to build their products to. The flawed assumption is that the ten or 15-year-old specification is still valid and accurate. This risk may occur during *modification* processes when obsolete modules are replaced with later versions.
- Technical optimism. Technical risks may appear when optimistic estimates are made about performance. A common example of this is the estimate of aircraft weight. Although efforts are made to reduce weight as time goes on, overweight on first delivery can be costly to the developer.
- Lack of internal analytical capability. This internal deficiency can result from the lack of analytical capability. For example, the developer may not have the capability to conduct the analysis of complex aerodynamics using a computational fluid dynamics (CFD) simulation or a structural analysis using a finite element analysis (FEA). Without this capability the developer may not be able to predict problems in these areas. Without structural analysis the developer may not be able to predict the physical interference between structural elements, such as control surfaces and the wing. These elements may flex substantially during flight depending on the speed, altitude, and amount of fuel on board.
- Lack of production quality. Production quality tends to fall due to the turnover of production workers. This deficiency can occur when one aircraft production comes to an end and workers are transferred to another aircraft. In these cases production quality can only be maintained if the new workers are given training and detailed instructions. A worker, for

example, may place two electrical cables too close to one another causing electromagnetic interference (EMI) problems.

- Operator actions. The aircraft operator, that is, the airline company, may take actions that will result in risks. For example the airline company may deviate from the maintenance manual produced by the developer. This was the case with the 1979 Chicago O'Hare crash as explained by Reason (1997, p. 88).
- Lack of rigor. Failure to follow processes with rigor is a source of risk itself. This book is a catalogue of such processes that may be the source of a risk if not followed rigorously. A simple failure to specify the environment of a supplier product may lead to catastrophic consequences.
- The airline company may acquire components that are not flight qualified, that is, they do not meet the requirements of the components that were specified by the developer. They can either purchase these components, or they can lease them as explained in Section 2.6.

Warning signs

One of the most important aspects of risk management is the recognition of early warning signs. In the interest of getting the aircraft built and delivered fast, these warning signs are often ignored with predictable consequences.

- One of the most visible warning signs is test failures. These failures can happen months in advance of delivery to the developer. These failures are both a call to action and a sign of a high-risk product.
- Another warning sign is doing things that have been done before with negative consequences. The overweight aircraft is a good example. This is a warning sign that should be addressed during early advanced design.
- The next category can only be called a logical warning sign. If the electronic experts say that the module will take three years to develop, but you schedule a two-year development, then you know you have a risk.
- The next category is just supplier quality. If a supplier has a record of late delivery and poor performance, and you plan to use this supplier again, then there is a risk. If, on the other hand, you are forced to use this supplier because, for example, no one else makes the product, or you are compelled to use this supplier for other reasons, then there is a risk. In this case, this becomes an external root cause. Either way it is a risk.

Unknown unknowns

The question often arises: what if you don't know the source of a risk? A serious consequence occurs, and there were no warning signs. Such was the case with TWA 800. In 1996 the fuel tank of a 747 exploded killing all 230 people on board. This accident is discussed by Jackson (2010, pp. 68–69). The flammability of aircraft fuel has always been well-known. However, identifying all ignition sources had been difficult. In 1996 an explosion occurred as a result of a worn electrical wire that came into contact with the vapor in the fuel. So the question is: how can explosions be prevented regardless of the ignition source (one of the unknown unknowns)?

The Federal Aviation Administration (FAA) (2008) addressed this problem by mandating that the vapor area in the fuel tank be filled with either nitrogen enriched air (NEA) or foam. This was a mitigation step that was independent of the source of the risk or the probability of its occurrence. The lesson of this example is that if a risk can be mitigated that is independent of the source that method should be employed or at least be considered a high-priority solution.

Off-the-table risks

Developers are sometimes reluctant even to record risks of certain types. Here are a few examples:

- If the risk source is a customer generated, the developer may not want to record the risk. For example, if the customer has a poor record of safety, maintenance, or pilot training the risk may not be recorded. Among options for handling this risk, the developer may want to discuss these items privately with the customer.
- If the risk is considered a *normal* problem, that is something that occurs all the time, the developer may consider it not worth recording. For example, if a supplier has a bad performance record, rather than recording it as a risk, the developer may chose to deal with this problem in the way that it is usually handled. However, if the usual way is not satisfactory, there is no reason for ignoring it as a risk.
- If the risk is considered *unsolvable* or the cost is great, the developer may not record it. For example, if the solution requires reversing a major design decision, this risk may go by the wayside. On the other hand, if the risk is recognized, then the various ways of handling it can be considered.

The lesson learned from the above examples is that risks should be recorded and treated whether they are considered normal or unsolvable.

15.4   Risk Mitigation Steps

Mitigation steps are the actions taken before a consequence is realized to lessen the magnitude of the consequence or the probability of occurrence. Mitigation steps will depend on the type of risk being mitigated. Most mitigation steps will cost money, but often the cost is worth the amount.

The simple test for how much money is too much is that the cost should be less than the cost of the expected consequence. If the cost of mitigation is more than the cost of the expected consequence, then the appropriate management method is risk acceptance.

However, here are a few examples of mitigation steps:

- Schedule slip. The obvious mitigation step for a schedule risk is to slip the schedule and suffer any costs associated with it. A tempting mitigation step is to increase the work load to solve the problem and get it done on time. However, this method is more likely to increase risk rather than mitigate it.
- Redesign. This can often be the most painful mitigation step. However, if it is clear that the current design will not meet the requirements, then there may be little choice. Metal fatigue is one example. Replacing the entire metal of the fuselage will be very expensive and will also increase the weight. Often the use of stiffeners will do the job.
- Change suppliers. If caught early enough, this risk can be mitigated will little cost and schedule impact. However, if the contract is already signed, this step can wind up costing money.

15.5 Issues

Consequences that have already been realized are not risks: they are called issues. For example, if cracks appear in the fuselage as a result of metal fatigue, this is an issue. The mitigation steps listed above will still have to be performed; the difference is that they are no longer optional; they have to be done.

Issues also exist if a consequence is inevitable. Consequences are generally only inevitable when the circumstances dictate a solution that is not optional. For example, an airline customer can dictate a solution that the developer knows is not feasible without consequences. The developer must take action to mitigate the consequences.

There is a common belief that all risks for which the likelihood is 1.0 are issues. This is not always the case. It is only the case for which the consequence will occur if there is no mitigation step to prevent it. Take the case of the component development schedule that was badly estimated. The consequence will only occur if the schedule is not changed. So the schedule change becomes the mitigation step.

15.6 Independent Review

One of the risks within the risk process itself is the fact that the risk may not be impartially identified, analyzed, or managed. There is an abundance of literature that supports this assertion. Vaughn, for example, analyses the NASA culture leading up to the *Challenger* disaster. Her conclusion was that risk was "normative," that is to say risks were just considered normal and were not treated with any special attention. Similarly, the Columbia Accident Investigation Report (2003, p. 189) concluded that NASA had a "broken safety culture." This book is not going to provide a formula on how to change the culture of an organization. Many methods have been proposed on how to do that. Jackson (2010, pp. 106–112) provides a summary of frequently suggested methods.

An independent authority

Rather, we will focus on a more agreed-to approach, the appointment of an independent authority to review risks, design decisions, and any other issues that may arise. The Columbia Accident Investigation Report (2003, p. 227) recommended an Independent Technical Authority. Similarly, following the 2006 crash of the UK aircraft the *Nimrod*, a derivative of the *Comet* aircraft, the Haddon-Cave report (2009, p. 393–495) recommended a rigorous airworthiness process. Haddon-Cave himself cited the Columbia recommendation as a model for independent review.

Independence

So having raised the topic of independent review, we have to ask, what is independence? In simple terms,

> Independence is the absence of any financial or organizational links between the reviewer and the project being reviewed.

So having defined independence, we have to ask whether independence is even possible. After all, we all work for the same government or the same enterprise. Nevertheless, some practical guidelines are possible to establish:

- An independent reviewer cannot report to the same program manager as the project being reviewed.
- Preferred reviewers report to a completely different program or if possible a completely different enterprise.
- Outside institutions, such as universities, are sometimes sources of qualified reviewers.

Regulatory agencies

Regulatory agencies, such as the FAA, are ideal independent bodies to a certain extent. However, they are generally limited to the review of safety issues. If a commercial aircraft company makes an unwise marketing decision, the regulatory agency has little authority to speak to this issue. Hence, the onus is on the company to implement the independent review process.

As discussed in Section 10.4, the Commercial Aviation Safety Team (CAST) of which the FAA is a member has recommended review by "third parties" of pilot training. This recommendation is based on the observation that pilots sometimes fail to execute their tasks in the manner they were trained.

Creating the independent review process

Creating the independent review process is no easy task. There are manifold obstacles. Yet it is of paramount importance. The main obstacle is that no program manager wants to yield any authority to an outside person. Hence, this authority must be yielded *voluntarily*. At a very minimum the program manager must be willing to accept the advice of an outside reviewer.

Sometimes the independent review may have some pretty serious and costly advice. For example, they might recommend more testing or even a major change in design or supplier. This book suggests the implementation of the independent review process may be the most important and the most difficult to implement. Yet the failure to implement it may have serious consequences.

15.7   The Risk Management Process

In the commercial aircraft domain it is unlikely that there would be a dedicated risk management organization. Risk management, as a process, would focus on the SE lead who would report directly to the program manager. Some of the responsibilities of the risk management lead are as follows:

- Create a risk management plan that would lay out how risk management is to be executed and what the roles and responsibilities are for the participants. This plan would explain the independent review process described above.
- The risk plan above would explain who the risk owner is, that is, the person responsible for explaining how each risk would be handled.
- Maintain a risk register showing the status of each risk and the handling method being employed.

15.8   Risk Management Tools

Many, but not all, commercial aircraft companies have risk management tools. Even those that do sometimes fail to use them, even for the most egregious risks. Some tools are internally created, and others are commercially available. It is not the purpose here to list or evaluate those commercially available tools.

First, having a risk management tool is not the most important aspect of the risk management process. It is a convenient way to record the risks and make them available for later review. Risks can be recorded using standard word processing or spreadsheet applications. Risk diagrams can be, and often are, created using standard chart creation applications, such as PowerPoint. These methods can be somewhat labor intensive, but if nothing else is available, it can be done.

Some risk management tool features

Here is a short list of some of the desirable features of risk management tools. Not all existing risk management tools have all these features. However, given the risk management process described above, these features would seem essential.

- The capability of recording the root cause of a particular risk. This feature is sometimes called the source. The warning signs can be considered a part of this feature.
- The tool should be capable of recording all possible consequences of a particular risk. The three major categories are technical, cost, and schedule. These are the consequences that would occur *if nothing is done to prevent them from happening.*
- The tool should be capable of recording the likelihood that the risk will occur if nothing is done to prevent it. This likelihood will be the same for all three consequences since all the consequences apply to the same risk. In the ideal case these likelihoods can be estimated using the warning signs described above. For example, if a supplier has a history of delivering their product late 50 percent of the time, then the likelihood is 0.5.
- The tool should be capable of calculating a risk level for each of the risk categories.

15.9 Opportunities

Opportunities are the counterpart to risks. Opportunities are the potential benefits that a program may realize if the consequences are better than planned for. For example:

- wheel balancing tests may reach their desired goal in fewer than planned for tests;
- the aircraft may actually weigh less than the design weight.

So the question is: How will be aircraft developer take advantage of these opportunities? There may be cost or schedule benefits to be realized as a result of the uncertainties in the system. These uncertainties are the same uncertainties that might have resulted in unfavorable consequences in other circumstances. But in the alternative situations they may yield opportunities in cost or schedule, for example.

The risk process should track these opportunities as well as risks. In practice the opportunities will be less numerous than the risks. However, the aircraft developer should not fail to take advantage of them.

15.10 Challenges for Risk Management

The MIT-PMI-INCOSE Guide for Lean Enablers (2012, p. 31) provides a list of challenges for the implementation of risk management. They are as follows; these items are paraphrased for brevity:

- insufficient involvement of professionals in risk management;
- lack of understanding of program risks;
- insufficient resources and funding for risk management;
- neglect of human aspect of risk management. This item reflects the cultural barriers to risk management as reflected above. Among these is the failure of management to defer to expertise on risk and to accept the risks that have been identified, without modification or reduction;
- disconnect between risk management and other program management processes;
- failure to resolve risks quickly.

This is a concise list of challenges that management needs to address the risks that are common on commercial aircraft.

16

Resilience of the Aircraft System

A topic of increasing interest in SE is resilience. Both the *Systems Engineering Body of Knowledge* (SEBoK) edited by Pyster et al. (2012) and the *International Council on Systems Engineering (INCOSE) Handbook* (2006) have sections devoted to resilience.

Resilience differs from safety in that safety endeavors to prevent the failure of a system, in this case an aircraft. Resilience goes beyond safety in that it calls for mechanisms to anticipate failure and to enable the system to recover from a major disruption, such as human error or an encounter with an external threat such as a flock of birds.

In this chapter we will discuss the basic principles of resilience and show how they can be applied to the design of an aircraft.

16.1 The History of Resilience

The history of resilience begins with the basic definition of resilience and shows how this discipline has evolved into its present meaning as a field within the context of SE and as applied to commercial aircraft.

The classical definition of resilience from the *Oxford English Dictionary* (OED) (1973, p. 1807) is that resilience is "the act of rebounding or springing back." This definition described the inherent property of an entity, for example a spring, a person, or an ecosystem, to recover from a disruption.

In more recent years the emphasis on resilience has shifted to systems which can be *engineered* to bounce back, that is, some degree of human intervention is required to enable the system to bounce back. Since these systems include humans, the term *engineer* has to be interpreted in a broad way and not to just classical engineering which is a branch of the physical sciences. Resilience pioneers Hollnagel and Woods (2006) and Hollnagel et al (2011) have coined the term *resilience engineering* and have published several books on the subject. The second book has considerable emphasis on the resilience of commercial aircraft.

Another step in the evolution of resilience was the appearance of papers and books with an emphasis on whole systems, that is, systems that consist of humans and technological elements as well. Hollnagel and Woods place considerable emphasis on organizational systems, while later researchers, such as Haimes (2009), consider the resilience of the entire system including technological elements as well as the humans. This book has adopted the whole-systems perspective since aircraft have many technological elements that play a major role in resilience.

16.2   The Definition of Resilience

It is possible to find hundreds of definitions of resilience. However, it not necessary to review all of them since they tend to have a great deal in common. The definition by Haimes is one of the few to appear in peer-reviewed journals. It is as follows:

> Resilience is the ability of the system to withstand a major disruption within acceptable degradation parameters and to recover within an acceptable time and composite costs and risks. (Haimes 2009, p. 498)

This definition covers a great deal more territory than just bouncing back. It even includes some aspects of interest to systems engineers, such as cost and risk.

The disruption cycle

The definition of resilience is inextricably intertwined with the disruption diagram shown in Figure 16.1. This diagram depicts a system, for example an aircraft, which has encountered a threat, at a time called the *event*. The encounter of US Airways Flight 1549 as described by Pariès (2011, pp. 9–27) with a flock of geese is such an event. Before the event the aircraft will be at an initial state in a normal operating mode. Following the event the aircraft will enter a recovery state in which it is attempting to recover from the damage encountered from the threat. The final state is either a full restoration of the functionality of the aircraft or some other acceptable state. In the case of Flight 1549 the final state was the ditching of the aircraft in the Hudson River where it floated long enough for the passengers and crew to be rescued. The diagram also allows for multiple or serial threats. For example, if a human error should occur during the attempt to recover the aircraft that would be considered another threat. In the case of Flight 1549 there was no such human error.

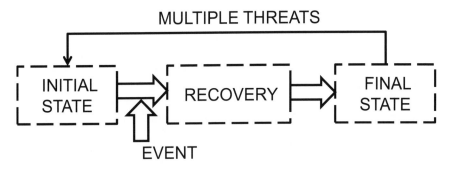

Figure 16.1   The resilience disruption cycle

Proactive vs. reactive perspectives

One issue that has dominated the discussion of definitions is whether resilience is proactive or reactive. Proactive resilience means that resilience covers the period of time before the encounter with a threat as shown as the initial state in Figure 16.1. Woods (2006, pp. 21–34) and Leveson et al. (2006, pp. 95–123) have adopted this perspective. In the proactive perspective more could be happening during the initial state of Figure 16.1; the pilot or the aircraft instrumentation could be detecting an approaching threat, for example, a flock of geese or a cloud of volcanic dust.

Reactive resilience means that resilience only applies during and after the encounter with the threat. Haimes (2009, pp. 498–501) adopted this perspective. Haimes refers to the period before the encounter as protection. The conclusion is that this distinction is one of definition and not of substance. Either perspective is valid providing analysts are consistent in their definitions, assumptions, and approach. This book has adopted the proactive perspective since this perspective is consistent with most of the literature consulted and is a simpler all-inclusive perspective.

16.3   Is Resilience Measureable?

As discussed by Jackson and Brtis (2015) whether resilience is measurable is the subject of debate. Haimes argues that it is not simply because there are too many dimensions to it; e.g., multiple threats, multiple failure modes and multiple recovery modes. However, it can be argued that while these complications make it hard to predict the resilience of a system, they do not preclude the identification of a metric, by which resilience can be gauged. This is an issue to be resolved in future research.

16.4   Design Rules and Example Solutions

This section will first outline a set of design rules that have been collected from various sources and then provide some example solutions that have been posed in the literature, especially those solutions that pertain to the commercial aircraft domain. These design rules are, first *abstract*, that is, they are simplified replicas of concrete solutions. Since these rules are abstract they can first be applied to any domain where the solutions seem logical and practical. Secondly, there may be many concrete solutions that can be derived from a single abstract rule. Figure 16.2 provides a simplified schematic of how an abstract rule relates to a concrete solution. In this case the abstract rule of *functional redundancy*, called *design diversity* by Leveson (1995, pp. 433–437), can be implemented as two separate aircraft control systems. The feature that abstract rules and concrete solutions have in common is a common set of dominant characteristics. The dominant characteristics for *functional redundancy* are that the system must consist of at least two branches, that these branches must be physically different, and that the branches must be independent.

Jackson and Ferris (2013, pp. 152–164) have compiled a list of these rules, which they call principles, and some associated support rules. Support rules basically have the same goals as the primary rules except that they have a more limited scope. These rules are sometimes called principles (by the author), characteristics (by Woods (2006, pp. 21–34)), or heuristics (by Rechtin (1991, p. 18)).

Since these rules are abstract, there is no way of knowing which rules are better than other rules. The analyst will need to model the solutions to determine that. In other cases the best solution will seem obvious. As the previous paragraph stated, any abstract rule may be transformed into many concrete solutions. It is therefore impossible to show all possible concrete solutions. The concrete solutions to follow will be extracted from the literature or posed as hypothetical examples.

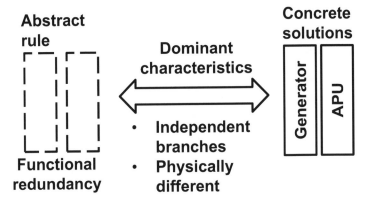

Figure 16.2   Abstract rules and concrete solutions

As we review these rules, though, it must be remembered that very rarely can the rules be implemented singly but must be implemented in combination with other rules. These combinations of rules result from the *interdependency* of the rules and the vulnerabilities of the individual rules. These interdependencies will be discussed within each section.

It is not the intent here to discuss all of the approximately 34 rules and support rules, only to provide examples for some of the more salient rules for the commercial aircraft domain.

Next, the word *design* should not be interpreted in the literal sense as in the design of hardware and software. It can imply the architectural arrangement of physical assets or of humans in the system. It can also imply the design of a system of procedures to be used by the humans in the system.

Finally, the reader should not interpret these rules to constitute a prescriptive process for designing a system. Rather, according to Hollnagel and Woods (2006, p. 348), "resilience engineering requires a constant monitoring of system performance, of how things are done."

The Absorption rule

Woods (2006, p. 23) refers to the *absorption* rule as the buffering characteristic. An obvious absorption capability of an aircraft is the design loads. One might ask: If an aircraft is designed with an adequate *absorption* capability, why would any other rules be required? The answer is that aircraft are designed knowing that abnormal conditions will occur that will exceed the design loads. Another answer is that even under normal conditions the design loads will be exceeded a small, but known percentage of the time.

Take the US Airways flight 1549 case discussed by Pariès (2011, pp. 9–27). All aircraft must meet stringent requirements for bird strike capability. The aircraft in question obviously met these requirements; however, on this occasion they were exceeded. To meet the resilience capabilities, the pilot was called into action using the *human in the loop* rule described below.

Another *absorption* capability is the lightning strike capability. The expected magnitude of lightning strike is 200,000 amps. This current is dispersed throughout the aircraft where it must pass through conduits without causing a conflagration. However, as the paragraph above stated, this load can be exceeded a small percentage of the time. To meet this requirement an adequate *margin* must be added to the design load.

The Limit Degradation Support rule

A pervasive threat to the *absorption* capability of an aircraft is the degradation of the capability due to poor maintenance or aging. This threat requires the invocation of the *limit degradation* support rule described by the Jackson and Ferris (2013, pp. 152–164).

Poor maintenance and aging are a constant threat to commercial aircraft. One of the more notable examples was the crash of the *Nimrod* aircraft which caught fire due to the failure of fuel seals as reported by the Royal Air Force Board of Inquiry (2007, p. 2–20).

The Margin Support rule

It is a common practice in almost all industries to add a margin to the absorption capability especially in the commercial aircraft domain. This margin allows for uncertainties in loads and in the capability of the aircraft. For most structural components the margin is 50 percent. Woods (2006, p. 23) lists margin a characteristic of resilience.

Dependency of the Absorption rule

As discussed above, most design rules should be employed in combination with other rules in order to be effective and to compensate for its vulnerabilities. In the case of *absorption*, this rule is dependent on both the *limit degradation* and *margin* support rules. The author (2013) discusses other dependent rules the designer may want to consider.

The Physical Redundancy rule

Physical redundancy, also called *design redundancy* by Leveson (1995, pp. 433–437), is a recognized concept in engineering. The dominant characteristics of physical redundancy are that the system should have two physically identical and independent branches. However, Rijpma (1997, pp. 15–23) points out that *physical redundancy* has more vulnerabilities than is generally recognized. For example, if an aircraft has two identical software systems, and one of these has an undetected flaw, the other system will have the same flaw.

Some aircraft employ the two-load-path design philosophy. This is a form of *physical redundancy*. That is, if part of the structure, for example the skin, fails due to metal fatigue or any other reason, the rest of the structure will be able to sustain the loads that are incurred.

Dependency of the Physical Redundancy rule

To compensate for the vulnerabilities of the *physical redundancy* rule, the designer can invoke the *functional redundancy* rule discussed below or call on the services of a person employing the *human in the loop* rule.

The Functional Redundancy rule

The *functional redundancy* rule, called *design diversity* by Leveson (1995, pp. 433–437), avoids most of the vulnerabilities of *physical redundancy*. The dominant characteristics are that the system must have at least two physically different and independent branches.

An aircraft having both mechanical and fly-by-wire (FBW) control systems is an example of *functional redundancy*.

No dependent rules were identified for the *functional redundancy* rule.

The Layered Defense rule

Layered defense is a derived rule from Reason's (1997, p. 11) Swiss cheese model. This model states that for a failure to occur a disturbance has to penetrate a series of layers similar to layers of Swiss cheese. In this model each layer has *holes*, that is vulnerabilities which allow the disturbance to travel to the next layer. So the implication of the model is that when a system has more layers of defense, the more resilient the system will be. Pariès (2011, pp. 9–27) calls this rule "defense in depth." Like *reduce complexity* and *reduce hidden interactions, layered defense* is one of the key rules that enables the system to maintain a distance from its safety boundary, that is, the increased number of layers enables it to stay farther from its safety boundary.

US Airways Flight 1549 described by Pariès is an example of *layered defense*. As described by Pariès, this system had four layers of defense. Two of these layers failed and the other two enabled the aircraft to ditch successfully. The first layer is the corrective measures to rid the airport area of birds that may harm the aircraft. The second layer is, of course, the engines themselves and their certified capability to handle a bird strike to a given capacity. Obviously, this capacity was exceeded and the engines failed. The third layer was the aircraft itself. Pariès points out the various features of the aircraft that enabled the pilot to maintain control even though the engines had failed. There was, for example, the APU (auxiliary power unit) that enabled the aircraft to provide electrical power. There was the RAT (ram air turbine) which provided hydraulic pressure. Then there was the flight control system that allowed the aircraft to function normally. The fourth layer was the pilot himself. It was his extensive training and psychological preparedness that allowed him to bring the aircraft in for a controlled ditching. This fourth layer was also an example of the *human in the loop* rule discussed below. This example also serves to illustrate the interdependency of the design rules.

Another example of *layered defense* is the prevention of collisions between aircraft in flight. The first layer belongs to air traffic control (ATC) who are responsible for tracking the aircraft by radar and making sure that they are at the assigned altitudes, which should be different for each aircraft. This layer uses the *drift correction* design rule. The second layer involves the use of the Traffic Collision Avoidance System (TCAS) described below. This system warns the pilot of an impending collision with another aircraft. Hence this layer also employs the *drift correction* design rule. However, a third layer is required in the form of action by the pilot to perform a corrective maneuver. Hence, the *human in the loop* design rule is required also.

Layered defense dependency

Layered defense does not itself have any dependent rules. However, layered defense is dependent on the implementation of at least two other rules. In the case of US Airways Flight 1549 there were four rules: (1) *drift correction* (corrective action to keep the birds away from the airport), (2) *absorption* (having engines with sufficient capacity), (3) *functional redundancy* (having alternative ways to control the aircraft), and (4) *human in the loop* (having a pilot with training and instincts to perform the task). As was stated above, it was the last two rules that enabled the aircraft to ditch satisfactorily.

The Human in the Loop rule

The *human in the loop* rule, discussed by Madni and Jackson (2009, pp. 181–191) is one of the most accepted rules in the commercial aviation domain. This rule asserts that humans should always be employed as system elements when there is a need for human cognition. The fact that human pilots continue to be the dominant mode of aircraft control attests to this assertion. Human dominance in ATC is further evidence of its validity.

The Automated Function Support rule

Billings (1991, pp. 244–245) elaborates on the *human in the loop* rule by saying that when there is a choice between the human or an automated system performing a given function, the task should always be performed by the human provided the task is within the time limitations and capability of the human.

The designer will need to balance this rule against the possible advantages of the *flight envelope protection* system discussed in Chapter 2. As discussed in that chapter, the purpose of that system is to prevent the pilot from allowing the aircraft to exceed its flight envelope.

The Reduce Human Error Support Principle

It is axiomatic that human error cannot be eliminated. However, there are numerous accepted ways to reduce and minimize human error. Reason (1990, pp. 217–250) presents a comprehensive list of these methods.

Human in the loop dependency

The *human in the loop* rule is primarily dependent on the *reduce human error* support principle to compensate for its primary vulnerability, that is, human error.

The Reduce Complexity rule

Complex systems are subject to highly erratic operation and sometimes failure. The *reduce complexity* rule states that the complexity of the system should be reduced as much as possible. Like *layered defense* and *reduce hidden interactions*, the purpose of this rule is to remove the system from its safety boundary as much as possible

The phenomenon of complexity is prevalent in both the aircraft system and the larger system called the large-scale supply chain system described in Chapter 14. One of the effects of complexity is the phenomenon of hidden interactions discussed in the section on the *reduce hidden interactions* rule discussed below. According to Marczyk (2012) there are three contributors to complexity as discussed below.

The first contributor is the number of elements of the system, in our case the number of parts of the aircraft. The number of elements by itself does not create complexity; however it is an important multiplier. It is possible for a system to have many components and not be complex. Such a system is called *complicated* rather than complex. The example often given, by for example Giachetti (2010, pp. 34–36), is a clock whose parts fit together perfectly. It is the lack of a perfect fit that makes the system complex; this lack of fit is called *variability* as discussed below.

The second contributor is the number of interfaces. Some authors, for example Carson (2000), consider this to be the primary contributor. With respect to the number of interfaces, this is an aspect entirely within the purview of the aircraft designer, in particular, the aircraft architect who determines the number and relationships among the subsystems. The methodology the designer might use is called cluster analysis as described by Hitchins (1993, pp. 135–147). The premise of cluster analysis is that the parts of the aircraft with the greatest *functional binding* will be a candidate to be an identifiable subsystem and will have the fewest interfaces among those parts that constitute the subsystem. As an example, the designer may want to decide whether to make guidance and navigation two software modules or one. Cluster analysis will help make that decision by minimizing the number of interfaces.

The third contributor is the variability in the relationships among the parts of the aircraft. Chapter 14 describes this factor more in depth. However, in short, this factor has to do with the variability in the quality of the data or other parameters that are exchanged among the parts of the aircraft. All of the parameters listed in Chapter 6 are candidates for variability.

The Reduce Variability Support Rule

As described above, variability is one of the primary contributors to complexity. Hence, reducing complexity requires reducing variability. Variability can be seen in its simplest technical terms, such as the variability of an electrical current between two components. If there are wide fluctuations in this current, then there will be variability and resultant complexity. Of course, real complexity is a composite of all the fluctuations of different parameters all over the aircraft. Chapter 14 also discusses the variability of parameters between the various elements of the supply chain also contributing to its complexity. Hence, rigorous management of all of these parameters will minimize the complexity of both the aircraft and the supply chain.

Reduce complexity dependency

There are two categories of rules upon which the *reduce complexity* rule is dependent. The first dependency is the *reorganization* rule described below. Reorganization includes both the reduction in the number of elements and the number of interfaces described above. The second category is the *reduce variability* rule described above.

The Reorganization rule

Woods (2006, p. 23) refers to this rule as the *restructuring* rule. He states that in order for a system to be resilient, it must be capable of restructuring itself. Although it may seem unreasonable for an aircraft to restructure itself during operation, it has happened.

The most obvious example is the Sioux City DC-10 crash of 1989. In this case, the aircraft lost control when its control system was damaged. The pilot managed to maintain some degree of control by using the propulsion controls. Most, but not all, of the passengers survived. This change of controls was tantamount to reorganizing the control system of the aircraft. As mentioned in Chapter 2, the FAA did not see fit to mandate propulsion control. Nevertheless, this incident did illustrate the principle of reorganization and its value.

Reorganization dependency

The *reorganization* rule is primarily dependent on the human in the loop rule because reorganization normally requires humans to implement the rule.

The Drift Correction rule

Drift correction, discussed by Dekker (2006, pp. 77–92) is the primary rule in the pre-event phase of disruption. It is the rule that enables the system either to anticipate or detect its drift towards an unsafe condition and makes a corrective action. *Drift* correction can either be in real-time or a long-term anticipation of a problem which may involve latent conditions not evident to the operators. Reason (1997, pp. 10–11) describes latent conditions in which flaws in the system are hidden until a catastrophic failure occurs.

The concept of latent flaws brings to mind a well-known axiom of system safety, paraphrased as follows:

> *Just because your aircraft has not had an accident does not mean it is safe.*

One of the most well-known threats to aircraft safety is bird strikes as discussed by Pariès (2011, pp. 10–11) that was a major factor in the US Airways Flight 1549 case. So the anticipation of this threat and steps to thwart it are an example of long-term drift correction. Skybrary (2013, unpaginated) points out that in many parts of the world mammals, such as reindeer, are also threats to aircraft because they can unexpectedly walk onto runways. Many corrective actions have been taken, such as reducing the bird population. However, as was seen in the Flight 1549 case, this was not effective, hence the need for *layered defense* as discussed above.

Typical of latent flaws in the aircraft domain is the degradation of the aircraft due to poor maintenance or aging as described above for the *Nimrod* case in the discussion of the *limit degradation* support rule. Corrective actions for these flaws are more frequent, and detailed inspections and oversight of the maintenance operations are required to detect them. For some types of flaws, for example structural cracks due to fatigue, special equipment may be necessary to detect them.

Regarding real-time *drift correction*, some existing methods already fall into that category, for example Terrain Avoidance Warning System (TAWS) as described by Skybrary (2012, unpaginated). This system warns of an approaching mountain or any other geological feature that may be a hazard to the aircraft and allows the pilot to perform a corrective maneuver.

Another example of *drift correction* is the Traffic Collision Avoidance System (TCAS). This is mandated system that uses a transponder to warn the pilot of the possible collision with another aircraft.

Another interesting example of real-time *drift correction* is the volcanic ash detector, called AVOID, which is in the planning stage, as reported by the BBC News (2010). This is a joint project between the airline EasyJet and the aircraft company Airbus. This device would warn the pilot if the volcanic ash density is becoming too dense and thus allow the pilot to take an alternative route. This development hinged on the adoption of a criterion for the maximum particle size of volcanic ash through which an aircraft could fly as also reported by the BBC (2010). This development followed the eruption of the Icelandic volcano Eyjafjallajökull in 2010, disrupting air traffic. It is not known whether this device would be mandatory on aircraft, but in any case, some airlines may voluntarily choose to install it.

Drift Correction Support rules

The *drift correction* rule has two important support rules: the *detection* support rule and the *corrective action* support rule. The *detection* support rule can be interpreted broadly. It can mean, for example, the detection of volcanic ash, as described above or the detection of hidden flaws in a software system. *Corrective action* can be any action to correct these problems.

Drift correction dependency

The drift correction rule is dependent on the two support rules listed above; it is also dependent on the *inter-node interaction* rule to convey the detection information to the control center of the system. This control center may be the pilot.

The Inter-Node Interaction rule

The *inter-node interaction* rule is an adaptation of the *cross-scale interaction* concept described by Woods (2006, p. 23). This rule finds its roots in systems theory in which a basic property of a system is *cohesion* as described by Hitchins (1993, p. 55). That is to say, for a system to be a system—and an aircraft is a system—there must be a relationship among all the parts. We have seen much evidence of the relationships among the parts of the aircraft in all the interfaces discussed in Chapter 6. Of primary interest is the relationship between the pilot and the aircraft automation. Billings (1997, pp. 232–262) lays out his set of rules (which he calls *requirements*) for all of these relationships. All of Billings' rules are in fact *heuristics*, that is, rules based on his own experience and observations.

The Informed Operator Support rule

The *informed operator* support rule as formulated by Billings (1997, p. 240) states that the operator, that is the pilot, should be completely knowledgeable about the operations of the automated system. In other words, the automated system should not perform any operation that the pilot does not understand. This support rule is particularly important in the implementation of *flight envelope protection*, as described in Chapter 2. Billings argues that this rule is most important when the aircraft is not operating in normal operating conditions.

The value of this rule is evident in those cases in which the failure to observe it resulted in a catastrophic event. For example, according to Zarboutis and Wright (2006, pp. 359–368), the aircraft in the Nagoya incident of 1994 was in a go-around mode while the pilot desired to land. This miscommunication between the pilot and the automated system led to the ultimate crash and loss of life.

The Knowledge between Nodes Support rule

Another Billings (1997, p. 243) rule, the *knowledge between nodes* support rule is similar to the *informed operator* support rule except broader in scope. This rule pertains to any combination of elements on the aircraft, not just the pilot and the automated system. However, the Nagoya case study satisfies this rule as well.

The Human Monitoring Support rule

This rule is the reverse of the *informed operator* rule; it states, according to Billing (1997, p. 240) that the automated system should monitor the human pilot. According to Billings, one of the most important aspects of human monitoring that the automated system should perform is to know when the human has made an incorrect data entry.

The Automated System Monitoring Support rule

This rule, according to Billings (1997, p. 243) goes beyond the *informed operator* rule; it states that the human should know the *intent* of automated system actions. Once again, this rule seems to have been absent in the Nagoya incident. If the automated system does not behave in accordance with pilot intentions, this rule will have been violated.

The Inter-Node Impediment Support rule

This support rule can be inferred from various case studies suggest this support rule which was documented by Jackson and Ferris (2013, pp. 152–164). It says that there should be no administrative or technical impediments to communication or cooperation among the nodes of a system. Among the more notable examples of when these impediments existed was the 9/11 Commission's (2004, p. 11) concern that the FAA could not communicate directly with the pilot of United 93 that a terrorist threat was imminent. This lack of communication was the result of protocols in the FAA at that time.

The Reduce Hidden Interactions rule

Perrow (1999, pp. 79–86) states that many failures result from hidden interactions that result from excessive complexity. It is from this observation that the *reduce hidden interactions* rule is inferred.

There are many ways that interactions can occur on aircraft and cause failures. Prominent among these is the effects of electromagnetic interference (EMI) when electrical devices, such as solenoids or generators, are located near data lines.

A dramatic example of hidden interactions was the Helios 522 aircraft in which a malfunctioning pressurization system and the automated flight control system combined to cause the crew and all the passengers to die of hypoxic hypoxia as described by Dekker et al. (2008).

So how can hidden interactions be reduced? At first, it might seem like a formidable task, and it can be. The most tedious method would be a mapping of all possible interactions as suggested by Jackson (2010, p. 39). This method can be expedited by exploring specific categories of interactions one at a time, such as EMI.

A more global approach is to reduce the complexity of the entire aircraft in accordance with the reduce complexity rule discussed above. However, even this approach would involve examining the variability of the parameters between the individual elements.

Finally, Leveson et al. (2006, p. 97) suggests that the problem and the solution might be managerial in nature. That is to say, closer cooperation between different design groups would reduce the chance that there would be any inconsistencies in the design approaches that would cause hidden interactions. Within SE such a managerial approach exists; it is called integrated product teams (IPTs) as described in Chapter 12. As Chapter 12 explains, an IPT has members from all technical disciplines working on the same product, for example, the wing or fuselage. The premise is that this close working relationship would result in fewer inconsistencies and hence hidden interactions.

16.5   Other Rules

If readers have access to Jackson and Ferris's paper (2013, pp. 152–164), they will notice that there are rules other than the ones listed above. These other rules may have some application to the commercial aircraft domain, but perhaps in a more limited way. For this reason, the following list will briefly summarize these rules and some potential applications.

The Repairability rule

There is no doubt that aircraft need to be repaired from time to time. Chapter 5 has devoted some space to the subject of maintainability. With respect to resilience the type of maintenance that is probably of interest is field repairs, such as, damage due to volcanic ash.

The Localized Capacity rule

Localized capacity has to do with the localization of elements of the aircraft so that a failure in one does not cause a failure in the other. For example, one would not like a fault to cause a failure of all the engines.

The Loose Coupling rule

Loose coupling has to do with slack between elements of the system, so that a failure in one element does not propagate to other elements. In commercial aircraft probably the most applicable area of study is the cluster analysis described by Hitchins (1993, pp. 135–147). This analysis assures that subsystems are internally functionally bound, while connections between subsystems are loosely coupled. For example, there would be very little, if any coupling between the environmental control system (ECS) and the flight control system.

The Neutral State rule

Neutral state is a rule of interest mostly to pilots and to control systems. This rule allows the aircraft following a disturbance to enter a neutral state to allow the pilot to make important decisions.

16.6   A Final Word on Interdependency

The reader should not underestimate the importance of interdependency which is discussed above within the discussion of each design rule. The reason is that if the designer employs only the design rules one at a time without their dependencies, the resultant design will most likely be brittle, that is, not resilient.

The second thing that the reader might notice is that the design rules form a chain, that is, when design rule A is dependent on design rule B, and design rule B is dependent on design rule C, a chain of design rules will form that will comprise a whole system of design rules. It is this whole system the designer will want to implement and not just the individual rules.

Final Comments

In the first edition I tried to emphasize the common sense approach to systems engineering and in particular its application to the commercial aircraft domain. I still believe that common sense is important, but there is more to SE than common sense. It is a question of mindset; mindset is the collection of assumptions and philosophies that rule your life.

The Systems Mindset

A mindset of importance here is the systems mindset. If your education is in aerodynamics, for example, you may view the whole world in terms of Mach number or Reynolds number. (This was my undergraduate and master's field of study.) As time goes on, you may have a broader view of a system. It may even result in the realization that the aircraft is a system. Sometime later you may conclude that this system is not really complete until it includes the pilot. Your view of a system will get broader and broader. It may eventually include the airline, the regulators, and the passengers. My point is that once you have the systems mindset, then figuring out what to do next is almost automatic. Your job is to make sure the whole thing, the system, hangs together, works together, and performs the desired functions.

The Risk Mindset

The subject of risk permeates this entire book even though only Chapter 15 treats it as a separate subject. The point is that if the principles of SE are not adhered to, at least as adapted in this book, there will always be a risk of the system failing to meet its objectives or failing in its entirety. The risk mindset is a positive mindset that helps you be aware of possible risks and things you can do to manage them as described in Chapter 15.

There is a negative counterpart to the risk mindset; it is sometimes called the risk denial mindset. The risk denial mindset is self-explanatory; it is the mindset that says that there are no risks or at least risks that cannot be handled until it happens. In my previous book (2010, pp. 91–119) I made a collection of these negative mindsets which I called paradigms. Prominent among these is a famous one called the Titanic Effect which is the belief that your system is safe when it is not. The Titanic Effect is tantamount to risk denial. In my opinion, and there is abundant evidence to support the existence of this mindset. That is, there is a widespread cultural disorder of risk denial. *Challenger, Columbia, Concorde, Katrina,* and *Deepwater Horizon* are all products of risk denial.

The problem is that there are no straightforward solutions to risk denial. Telling decision makers to shape up is not a solution since they are usually the victims of risk denial. As explained in Chapter 15, the only concrete solution on the table is an independent authority which was recommended by the Columbia Accident Investigation Board. This solution is not easy either since there is little agreement on what independent means and how to achieve independence. In commercial aircraft the regulators supply a certain degree of independence. In the end, however, the aircraft developers need to embrace the concept of independence and implement it.

The Resilience Mindset

Resilience is a new concept that is gaining favor in many domains including commercial aircraft. As explained in Chapter 16, resilience goes beyond safety in that safety is concerned with preventing failures, while resilience deals with methods to anticipate, withstand, and recover from disruptions of many kinds. However, resilience requires a mindset of its own, which may be difficult for many engineers to adopt. As pointed out in Chapter 16, many authorities claim that resilience is , at the very least, difficult to measure. This is a fact that will be difficult for many engineers to accept. The net result is that in order to find a satisfactory solution, the designer will have to use an iterative, that is, trial and error solution.

Also, as pointed out in Chapter 16, many of the resilience concepts, such as *layered defense*, are already present, to some extent, in the commercial aircraft domain. The "Miracle on the Hudson" case illustrates that point. Nevertheless, acceptance of resilience principles and the adoption of the resilience mindset promise to make safer skies.

Appendix 1
The Mathematics of Reliability Allocation

The purpose of this appendix is to provide an explanation of how multiplicative parameters, such as reliability, are allocated. In this appendix we are using the third definition (see Glossary) of the word *allocation*, namely, a breakdown of a top-level requirements parameter into its component parts.

A1.1 Basic Reliability

The basic reliability equation is:

$$R = R_1 \times R_2 \ldots R_i!$$

What this equation says is that, if several components have reliabilities R1 through R_i, then the reliability of all the components working together is R. For example, for the case of only two components, if component 1 has a reliability of 0.99 and component 2 has a reliability of 0.99, then the reliability of the whole subsystem (consisting of the two components) is about 0.98.

A1.2 Allocation for Generically Similar Components

Allocation does it in reverse: that is, it sets a required value of R and determines what R_1 and R_2 must be to keep the reliability at least equal to R (that is, "allocates" values to R_1 and R_2). For example, let's say that R must be at least 0.99. Then one possible allocation is R_1 = 0.995 and R_2 = 0.995. (This can be checked by multiplying them together.) This would be a good allocation if both components were (1) of the same generic class (mechanical or electrical) and (2) these reliabilities were considered to be achievable. In our specification the allocation would look like:

Total reliability	0.99
Component 1	0.995
Component 2	0.995

A1.3   Allocation for Generically Different Components

Another possible solution occurs if the reliability specialist knows, for example, that the components are different: one is mechanical and the other is electrical. He or she knows that electrical components are more reliable than mechanical; the reverse is true for some types, but this is only an example. The first step is to set R_1 (electrical component) to 0.999 and R_2 (mechanical component) to 0.991. The resulting solution yields an acceptable value of $R = 0.99$. The table would look like

Total reliability	0.99
Component 1	0.999
Component 2	0.991

A1.4   Redundancy

Redundancy becomes necessary when the reliability specialist knows that the total reliability of 0.99 is not achievable with the available components. Let's say that the mechanical component still has a reliability (R_2) of 0.991. However the electrical component has a reliability (R_1) of 0.99. The net reliability (R) would then be only 0.981 (by the equation, above). This is obviously too low. What can be done? We can use redundant electrical parts, that is, if one of the parts fails, the other will still do the job. The equation for redundancy is a little more complicated:

$$R_1 = 1 - (1 - R_e)^2$$

where R_e is the reliability of an individual electrical part and R_1 is the reliability of the two redundant electrical parts working together. Thus in this case, if Re is 0.99, then R_1 would be 0.9999. Much better, right? Hence, the total reliability of the subsystem would be about 0.991, that is $R_1 \times R_2$, and we would have met the requirement of at least 0.99. The allocation table would look like:

Total reliability	0.99
Electrical assy	0.9999
Component 2 (mechanical)	0.995

Actually, in this case the achievable reliability will be greater than 0.99, namely, about 0.995 since the electrical assembly is now so reliable due to redundancy. However, the requirement will still be met. The reliability specialist may now choose to reduce the mechanical reliability allocation to 0.99 if there are some major cost savings to be gained.

A1.5 The Whole Airplane

Of course, our examples above are for only two components. In reality there may be thousands of components. The reliability specialist approaches this problem in the same hierarchical way as the weight specialist does, that is, by product center. The equation is a little longer, but not much, for example:

$$R_{airplane} = R_{wing} \times R_{fuselage} \times R_{empennage} \times R_{horiz} \times R_{systems} \times R_{propulsion}$$

The specialist then proceeds as shown in the above examples for only two components.

Appendix 2
Example Commercial Specification Outline

This example specification is adapted from MIL-STD-961D (1995), tailored, and simplified for commercial applications. For many years, the standard specification format was MIL-STD-490A (1985) which was recently superseded by MIL-STD-961D. While somewhat differently organized, these two documents cover substantially the same information. Military considerations, such as vulnerability, have been deleted. Terminology, such as *logistics*, not often used in commercial practice has been replaced with terms, such as *support*, more commonly used in commercial practice.

The reader may notice some differences in terminology between this example specification and the body of this book. For example, MIL-STD-961D places reliability and other constraints in the functional and performance requirements category rather than in a separate constraints category. These differences reflect some of the differences in theory and terminology among systems engineers. However, these differences are not significant from an implementation point of view: Most systems engineers are in basic agreement regarding the factors which should be considered.

This format cannot be considered rigid, but rather adaptable to the individual system being developed. This document can be used, with modification, for aircraft system level and aircraft-level and subsystem-level requirements in accordance with the principles of Chapter 4. This format can be used, at management discretion, first, as an airline customer document to clarify the requirements of the aircraft which the aircraft will be designed to meet. Secondly, it can be used as an internal document for the tracking and verification of requirements. Finally, the format can be used as a medium for establishing supplier requirements and conducting supplier management as explained in Section 12.9. This document can be part of the contractual documentation between the manufacturer and both the customer and the supplier.

A2.1   Scope

This section defines the scope of the document. It names the system (aircraft system, aircraft, or subsystem) being addressed and the purpose (establish requirements and verification methods) for that system. This section is purely introductory and does not contain any verifiable requirements. It describes the major components of the system, presents some diagrams, if applicable, and describes, in general terms, the external interfaces of the system. It describes any mandatory components, that is, non-development items (NDIs), which are particularly important for derivative aircraft as described in Section 2.2.

A2.2   Applicable Documents

A2.2.1   General

This section should list *all* the documents and *only* the documents which constitute part of this specification. These include company, industry, and regulatory documents, including applicable certification documents. These documents will be made available to the supplier if necessary.

A2.2.2   Certification basis

This section specifies the basis on which the aircraft will be certified as described in Chapter 10. Specific FARs will be cited.

A2.3   Requirements

All requirements in this section, with the exception of Sections A2.4.1 and A2.4.2, should be verifiable in accordance with the principles of Chapter 11 and should contain a *shall* as part of the requirements statement.

As explained in Section 4.8, various requirements may be in conflict. That section also explains several ways to resolve those conflicts. The results of that resolution should be documented in Section A2.8 below.

A2.4   Functional and Performance Requirements

A2.4.1   Missions

This section describes the missions of the aircraft or subsystem as necessary to describe requirements. These missions can include passenger or cargo missions as described in Section 3.3.

A2.4.2   Operational phases and modes

This section describes the operational phases and modes of the system, as described in Section 3.3. These phases or modes may differ for individual subsystems or components.

A2.4.3   Product capability

This is the primary paragraph in the specification. It should not be deleted without good reason. This is the paragraph that dictates how the product being delivered should perform.

All requirements in this section pertain to the performance of the system and must be derived from functions as described in Section 4.2. Each requirement should contain a *shall* to indicate that it is mandatory and must be verified. The term *entity* refers to the system being specified, whether at the aircraft system level, aircraft level, subsystem level, or component level. The term *capability* refers to performance requirements as defined in this book in Section 4.2.

A2.4.4   Reliability

This section specifies both the dispatch and operational reliabilities for the system as described in Section 5.4. It does not contain the safety related probabilities of failure discussed in Section A2.4.13 below.

A2.4.5   Maintainability

This section specifies the maintainability requirements on the aircraft as described in Section 5.7. It does not specify requirements for maintenance equipment as in Section A2.4.18 below, except for the specification for the support system as described in Section 2.3.

A2.4.6   Environmental conditions

This section describes all the natural and induced environments which the system must endure as discussed in Section 5.6. If the product is internal to the aircraft, the environment should reflect the section of the aircraft in which the product will be placed. These environments may vary widely.

If the product is external to the aircraft, this section should reflect that environment also.

A2.4.7   Transportability

This section specifies any transportability constraints on the system as discussed in Section 5.12.

A2.4.8   Materials and processes

This section specifies any limitations on materials, processes, and parts, as specified in company, industry, or government design standards as described in Section 5.8. Limitations specified by the customer may also be included. It may also specify any manufacturing processes which are needed to produce the item as described in Section 5.14.

A2.4.9   Electromagnetic radiation

This section specifies the electromagnetic radiation a component is permitted to *emit* as described in Section 5.10, as opposed to the electromagnetic environment of Section A2.4.6, above.

A2.4.10   Nameplates or product markings

This section specifies how the entity should be marked.

A2.4.11   Producibility

This section specifies any process requirements, such as tolerances, related to the manufacturing process as discussed in Section 5.14.

A2.4.12   Interchangeability

This section imposes interchangeability requirements on the design of the aircraft components as described in Section 5.7.

A2.4.13   Safety

This section imposes both quantitative safety requirements as described in Section 10.2 as well as the qualitative safety factors described in Section 5.5 and 10.2.

A2.4.14   Human engineering

This section specifies all requirements related to human factors as described in Section 5.5. This section is normally used in situations in which people are not part of the system, but interface with the system. This section is used in those situations to specify only the aspects of the system which may be affected by that interface. If people are part of the system, then specific requirements relating to people, such as, number, training, and physical characteristics, will be specified in Section A2.4.19.

A2.4.15   Security requirements

This section specifies any requirements related to the security of the system, either electronic or physical.

A2.4.16 Software requirements

This section specifies the top-level software requirements described in Section 10.3 Firmware requirements are also included in this section.

A2.4.17 Design and implementation constraints

A2.4.17.1 Physical characteristics
Physical characteristics include mass properties as described in Section 5.2 and dimensions as described in Section 5.3.

A2.4.17.2 Design-to-cost requirements
Although cost constraints are not covered in MIL-STD-961D, this section contains any cost constraints on the system as described in Section 5.11. Included are aircraft system-level costs as described in Section 8.1, Item 17, direct operating costs (DOC) described in Section 8.6, and indirect costs described in Section 8.6. These costs are normally not treated in military specifications.

A2.4.17.3 Noise
Noise is also not specifically mentioned in MIL-STD-961D. This section specifies the noise levels the system may emit as described in Section 5.9, as opposed to the noise environment of Section 5.6, above. This is an important constraint for commercial aircraft.

A2.4.17.4 Flexibility and expansion
In commercial aircraft, flexibility and expansion are important from a total aircraft point of view. MIL-STD-961D mentions flexibility and expansion in a software context. This section specifies features the aircraft should have in order to accommodate future growth options described in Section 5.13.

A2.4.18 Support requirements

This section specifies specific requirements for maintenance and other support. For the aircraft system, this section will simply refer to the support subsystem to be described in Section A2.9, below.

A2.4.19 Personnel and training

This section specifies the number and qualifications of personnel which will operate, maintain, or service the aircraft as described in Section 8.1. It also specifies the levels of training required for each personnel category.

A2.4.20   Requirements traceability

This section may consist of a table to provide traceability from each requirement to higher-level requirements, as discussed in Section 4.2.

A2.5   Interface Requirements

This section specifies the functional and physical interfaces between this entity and any other entity, as discussed in Chapter 6.

A2.6   Design and Construction

This section specifies design standards discussed in Section 5.8, workmanship requirements, and special production inspection requirements as required by the producibility requirements discussed in Section 5.14.

A2.7   Documentation

This section specifies any documentation which must be delivered as part of the deliverable package to the airline, such as training and maintenance manuals.

A2.8   Precedence of Requirements

The purpose of this section is to specify the order of relative importance of the requirements. For example, requirements related to safety would supersede any other requirements.

Section 4.8 explains how various requirements may be in conflict. That section also describes various ways of resolving those conflicts. This section should explain the resolution of that analysis and the resulting driving requirements.

A2.9   Major Component Characteristics

MIL-STD-961D does not have a section pertaining to the requirements of major component characteristics. However, some systems engineers find it useful. This section will provide the top-level requirements for the elements one level below the system addressed in this specification. For the aircraft system, these elements are the training equipment, support equipment, and facilities. For the aircraft level, these elements are the subsystems of Figure 2.1.

A2.10   Verification

This section contains the verification matrix and any other material related to the verification of requirements for the system as explained in Chapter 11. The verification matrix should contain all the elements required for certification discussed in Section 10.1 Section 10.17. It also contains the verification descriptions for all requirements.

A2.11   Preparation for Delivery

This section provides guidance on how specific elements of the aircraft are to be delivered to the manufacturer.

A2.12   Notes

Any information which should be made known as background information or as instructions to the suppliers may be included.

A2.13   Appendices

This section may be used for any information deemed useful. It may contain requirements which are of a temporary or limited nature, such as for a test aircraft or requirements for a specific customer. Only those requirements which are necessary to describe the temporary features of the item are listed in this appendix.

Appendix 3
Systems Engineering Automated Tools

Frequently SE is facilitated by an automated SE tool. In fact, the vast number of functions and requirements make automated tools a virtual necessity. This appendix summarizes the features and benefits of automated tools.

A3.1   Features of Automated SE Tools

Following is a summary of the basic features an automated SE tool usually has. The SE tool should be able to:

1. Support functional analysis.
2. Assign a requirement (or requirements) to each function as the requirements are developed. In some tools, the product of the requirements task is the requirements allocation sheets (RASs).
3. Allocate the requirements to specific components, subsystems, or systems. This allocation is also recorded on the RASs.
4. Create a traceability record of requirements so that the source of each requirement can be identified.
5. Sort for missing data in order to assure completeness.
6. Assign consistent design constraints in keeping with regulatory requirements and other sources.
7. Create specifications automatically from the data base in standard or custom specification formats.

Many tools have other specialized capabilities. These include configuration management and simulation capabilities. However, the above seven features are considered to be the most fundamental capabilities.

A3.2   Benefits of Automated SE Tools

Automated requirements tools have the following benefits:

1. They are fast and can accomplish the analysis in a small fraction of the time required by hand.
2. They provide the traceability of requirements discussed above.

3. They provide increased program technical control by assuring that all organizations, including suppliers, if required, are using the same data base. The tool becomes an integral part of the configuration management process.
4. They enhance data management by providing a single source of data.
5. They provide an efficient tool for fast change control.
6. They provide a medium for requirements data exchange among groups and with suppliers.

Bibliography

9/11 Commission. 9/11 Commission Report. Edited by T. H. Kean, 2004.

Adams, Charlotte. Shop Data Loaders for the Boeing 777, *Avionics Magazine*, September 1996, pp. 42, 43.

Airbus. Fly by Wire. Airbus [cited 26 December 2013]. Available from http://www.airbus.com/tools/airbusfor/pilots/fly-by-wire/?contentId=%5B_TABLE%3Att_content%3B_FIELD%3Auid%5D%2C&cHash=22935adfac92fcbbd4ba4e1441d13383, 2013.

Alexander, Christopher. *Notes on the Synthesis of Form*. Cambridge, MA, Harvard University Press, 1964.

ANSI/EIA 632, *Processes for the Engineering of a System*, Electronic Industries Alliance (EIA), Arlington, VA, 1999.

Army Field Manual. FM-770-78 Field Manual: *System Engineering*, Headquarters, Department of the Army, 1979.

Army Technical Manual. TM 38-760-1, *A Guide to System Engineering*, Department of the Army, TM 38-760-1, 1973.

Applegate, John. Systems Engineering in Developing Nations. INCOSE Symposium, 1998.

ATA Specification 100 - Specification for Manufacturers' Technical Data, Revision No. 37, Air Transport Association of America, 1999.

BBC. New Rules to Aid Ash Flight Chaos, 18 May 2010 [cited 31 May 2011]. Available from http://news.bbc.co.uk/2/hi/uk_news/8688517.stm, 2011.

BBC News. Easyjet to Trial Volcanic Ash Detection System, 10 June 2010 [cited 4 April 2011]. Available from http://www.bbc.co.uk/news/10234553, 2011.

Berry, Dennis L. Civil Aircraft Propulsion Integration—Present and Future, SAE Technical Paper Series, Number 932624, 27–30 September 1993.

Billings, Charles. *Aviation Automation: A Concept and Guidelines*. Moffett Field, CA: National Aeronautics and Space Administration (NASA), 1991.

Billings, Charles E. *Aviation Automation: The Search for a Human-Centered Approach*. Mahwah, NJ: Lawrence Erlbaum Associates, 1997, pp. 232–262.

Birch, Stuart. Technology Update, *Aerospace Engineering*, December 1995, pp. 9–10.

Bowers, Al. Blended-Wing-Body: Challenges for the 21st Century. NASA Dryden Flight Research Center 2000 [cited 9 January 2014]. Available from http://www.twitt.org/BWBBowers.html. 2014.

Buede, Dennis M. *The Engineering Design of Systems*. Edited by A. Sage, *Wiley Series in Systems Engineering*. Hoboken, NJ: John Wiley & Sons, Inc., 2000.

Campbell, CADRAT Tool. Personal communication. UK, 2013.

Carson, Ronald S. A Set Theory Model for Anomaly Handling in System Requirements Analysis, INCOSE Proceedings, 1995, pp. 515–522.

Carson, Ronald S. Global System Architecture Optimization: Quantifying System Complexity. In *International Council on Systems Engineering*: INCOSE, 2000.

Chapanis, Alphonse. *Human Factors in Systems Engineering*, New York: Wiley, 1996, pp. 14, 206, 207, 277.

Checkland, Peter. *Systems Thinking, Systems Practice*. New York: John Wiley & Sons. 1999.

Commercial Aviation Safety Team (CAST) 2011 [cited 16 February 2014]. Available from http://www.cast-safety.org/about_vmg.cfm, 2014.

Commercial Aviation Safety Team (CAST) Safer Skies Safety Enhancements for Manufacturers. 4 September 2012 [cited 15 February 2014]. Available from http://www.cast-safety.org/pdf/2012-09-04_Safer_Skies_Safety_Enhancements_for_Manufacturer.pdf, 2012.

Commercial Aviation Safety Team (CAST) Safer Skies Safety Enhancements Commercial Aviation Safety Team, 2014.

Conrow, Edmund H. *Effective Risk Management: Some Keys to Success*. Second edition. Reston, VA: American Institute of Aeronautics and Astronautics, 2003.

Corning, Gerald, *Supersonic and Subsonic CTOL and VTOL Airplane Design*. College Park, Maryland, published by author, 1977.

Defense Specifications, Department of Defense, MIL-STD-961D, 22 August 1995 (supersedes MIL-STD-490A).

Dekker, Sidney. Resilience Engineering: Chronicling the Emergence of Confused Consensus. In *Resilience Engineering*, edited by E. Hollnagel, D. D. Woods and N. Leveson. Aldershot, UK: Ashgate Publishing Limited, 2006.

Dekker, Sidney, Erik Hollnagel, David D. Woods, and Richard Cook. *Resilience Engineering: New Directions for Measuring and Maintaining Safety in Complex Systems*. Lund, Sweden: Lund University, 2008.

Department of Defense (DoD). Standard Practice: System Safety. Washington, DC: Department of Defense, 2012.

EasyJet. AVOID Volcanic Ash Detector. EasyJet [cited 12 June 2014]. Available from http://corporate.easyjet.com/corporate-responsibility/avoid-volcanic-ash-detection.aspx?sc_lang=en, 2014.

Electronic Industries Association (EIA), *System Engineering*, Engineering Bulletin SYSB-1, 1989.

Electronic Industries Alliance (EIA), Processes for the Engineering of a System, ANSI/EIA 632, January 1999.

Federal Air Regulation, Part 25, Airworthiness Standards, Transport Category: Airplanes; Federal Aviation Administration, Department of Transportation, March 1993.

Federal Aviation Administration (FAA). Fuel Tank Flammability Reduction Means. In *Advisory Circulars*, edited by Ali Bahrami. Washington DC: Federal Aviation Administration, 2008.

Federal Aviation Administration (FAA). Applicability/Compatibility of STPA with FAA Regulations and Guidance. Edited by Institute of Engineering and Technology. Seattle, WA: Federal Aviation Administration 2012.

Federal Aviation Administration (FAA). Fact Sheet—Commercial Aviation Safety Team. Federal Aviation Administration, 11 July 2013 [cited 15 February 2014]. Available from http://www.faa.gov/news/fact_sheets/news_story.cfm?newsId=15214, 2014.

Federal Aviation Administration (FAA). *FAA Systems Engineering Manual*. Edited by Kimberly Gill. Washington, DC: Federal Aviation Administration, 2014.

Generic Open Architecture (GOA) Framework, AS4893, Society of Automotive Engineers, January 1996.

Giachetti, Ronald E. *Design of Enterprise Systems: Theory, Architecture, and Methods*. Boca Raton, FL: CRC Press, 2010.

Grady, Jeffrey O. *System Requirements Analysis*, New York, McGraw-Hill, Inc. This book presents a comprehensive view of the development of requirements for a system,1993.

A Guide to System Engineering, TM 38-760-1, US Army, 1973.

Guidelines for the Certification of Highly-Integrated and Complex Aircraft Systems, Society of Automotive Engineers (SAE) in cooperation with the Federal Aviation Administration (FAA), ARP 4754, November 1996.

Haddon-Cave, Charles. *An Independent Review into the Broader Issues Surrounding the Loss of the RAF Nimrod MR2 Aircraft XV230 in Afganistan in 2006*. London: The House of Commons, 2009.

Haimes, Yacov Y. On the Definition of Resilience in Systems. *Risk Analysis* Vol. 29, No. 43, pp. 498–501, 2009.

Hall, Arthur D. *A Methodology for Systems Engineering*. Princeton, NJ: D. Van Nostrand Co., Inc., 1962.

Hamzeh, Osama N., W. Woytek Tworzydlo, and Hsien J. Chang. Analysis of Friction-Induced Instabilities in a Simplified Aircraft Brake. In *SAE 1999 Brake Colloquium*. San Diego, CA: Society of Automotive Engineers, 1999.

Hitchins, Derek. *Putting Systems to Work*. Hoboken, NJ: Wiley, 1993.

Hitchins, Derek. *Advanced Systems Thinking, Engineering, and Management*. Norwood, MA: Archtech House, 2003.

Hollnagel, Erik, Jean Pariès, David D. Woods, and John Wreathhall, eds. *Resilience Engineering in Practice: A Guidebook*. Edited by E. Hollnagel, S. Dekker, C. P. Nemeth and Y. Fujita, *Studies in Resilience Engineering*. Farnham, Surrey, UK: Ashgate Publishing Limited, 2011.

Hollnagel, Erik and David D. Woods. Epilogue: Resilience Engineering Precepts. In *Resilience Engineering: Concepts and Precepts*, edited by E. Hollnagel, D. D. Woods and N. Leveson. Farnham, UK: Ashgate Publishing Limited, 2006.

Hollnagel, Erik, David D. Woods, and Nancy Leveson, eds. *Resilience Engineering: Concepts and Precepts*. Aldershot, UK: Ashgate Publishing Limited, 2006.

Holy Bible, revised standard version, New York: Thomas Nelson and Sons, 1952, The New Testament, pp. 195–196.

Honour, Eric C. Requirements Management Cost/Benefit Selection Criteria, *Proceedings of NCOSE*, 1994.

Hooks, Ivy, and Kristan A. Farry. *Customer-Centered Products: Creating Successful Products Through Smart Requirements Management*. New York: American Management Association, 2001.

Hopkin, Paul. *Holistic Risk Management in Practice*. Livingston, UK: Witherby & Co, Ltd, 2002.

Hypersonic Aircraft Propulsion. *Aerospace Engineering*, June 1996.

IEEE Guide for Developing System Requirements Specifications, IEEE Std 1233-1996, 6 June 1996.

IEEE Standard for Application and Management of the Systems Engineering Process, IEEE Std 1220-2005.

INCOSE. *Systems Engineering Handbook*. Edited by SE Handbook WG. Seattle, WA: International Council on Systems Engineering, 2010.

INCOSE Fellows. A Consensus of the INCOSE Fellows. International Council on Systems Engineering [cited 2 June 2006]. Available from http://www.incose.org/practice/fellowsconsensus.aspx, 2006.

Jackson, Scott. Systems Engineering and the Bottom Line, INCOSE Proceedings, 1995.

Jackson, Scott. Introducing Systems Engineering into a Traditionally Commercial Organization, INCOSE Proceedings, 1996.

Jackson, Scott. *Systems Engineering for Commercial Aircraft*. Aldershot, UK: Ashgate Publishing Limited (in English and Chinese), 1997.

Jackson, Scott. *Architecting Resilient Systems: Accident Avoidance and Survival and Recovery from Disruptions*. Edited by A. P. Sage, *Wiley Series in Systems Engineering and Management*. Hoboken, NJ, USA: John Wiley & Sons, 2010.

Jackson, S. and Brtis, J. Overview of Resilience and Theme Issue on the Resilience of Systems. *Insight*, 18 (2015, April).

Jackson, Scott, and Timothy Ferris. Resilience Principles for Engineered Systems. *Systems Engineering* Vol. 16, No. 2, pp. 152–164, 2013.

Jamshidi, M. Systems of Systems Engineering: Innovations for the 21st Century. In *System of Systems Engineering: Innovation for the 21st Century*, edited by M. Jamshidi. Hoboken, NJ: John Wiley & Sons, 2009.

Kehlet, Alan. Major Advances in Aircraft Technologies Expected in the Future, *Innovate* bulletin, McDonnell Douglas, Vol. 26, No. 4, 4th Quarter, 1995.

Kossiakoff, Alexander, and William N. Sweet. *Systems Engineering: Principles and Practice*. Edited by Andrew Sage, *Wiley Series in Systems Engineering and Management*. Hoboken, NJ: John Wiley & Sons. 2003.

Lano, Robert J. *Techniques for Software and System Design*. Vol. 3, *TRW Series on Software Technology*. Amsterdam: North-Holland Publishing Co., 1979.

Leveson, Nancy. *Safeware: System Safety and Computers*. Reading, MA: Addison Wesley, 1995.

Leveson, Nancy. *A New Approach to System Safety Engineering*. Cambridge, MA: Massachusetts Institute of Technology, 2002.

Leveson, Nancy, Nicolas Dulac, David Zipkin, Cutcher-Gershenfeld, John Carroll, and Berry Barrett. Engineering Resilience into a Safety-Critical System. In *Resilience Engineering: Concepts and Precepts*, edited by E. Hollnagel, D. D. Woods and N. Leveson. Aldershot, UK: Ashgate Publishing Limited, 2006.

Lin, Kuen, Eric Cheung, Wendy Liu, and Luke Richard. Disbond/Delamination Arrest Features in Aircraft Composite Structures. In *2013 Technical Review*: Joint Advanced Materials & Structures Center of Excellence, 2013.

Littlewood, Bev, and Stringini, Lorenzo, The Risks of Software, *Scientific American*, November 1992, pp. 62–75.

Mackey, Dr William F. Conducting a Technology Management Assessment, INCOSE Proceedings, 1996.

Madni, Azad and Scott Jackson. Towards a Conceptual Framework for Resilience Engineering. *Institute of Electrical and Electronics Engineers (IEEE) Systems Journal* Vol. 3, No. 2, pp. 181–191, 2009.

Martínez-Val, Rodrigo, E. Pérez, T. Muñoz, and Cristina Cuerno. Design Constraints in the Payload-Range Diagram of Ultrahigh Capability Transport Airplanes, *Journal of Aircraft* Vol. 31, No. 6, November–December 1994.

Marczyk, Jacek. *Practical Complexity Management*. Trento, Italy: Editrice/UNI Service, 2009.

Marczyk, Jacek. Complexity Reduction. Como, Italy, 14 June 2012.

MIL-STD-499B. *Systems Engineering*, Washington DC: Department of Defense, 1994 (cancelled draft).

MIL-STD-1808B. *Interface Standard System Subsystem Sub-subsystem Numbering*, Department of Defense, 1 August 2007, Washington, DC.

NASA. *Concept of Operations for Commercial and Business Aircraft Synthetic Vision Systems*. Edited by Daniel M. Williams. Hampton, VA: National Aeronautics and Space Administration, 2001.

NASA. *Concept of Operations for Commercial and Business Aircraft Synthetic Vision Systems*, Version 1.0, NASA/TM-2001-211058, Langley Research Center, Hampton, VA, 2001.

NASA. *Columbia Accident Investigation Report*. Washington, DC: National Aeronautics and Space Administration (NASA), 2003.

NASA. Technology Readiness Levels. NASA 2012 [cited 2 March 2014]. Available from http://www.nasa.gov/content/technology-readiness-level/, 2014.

NASA Systems Engineering Handbook, SP-6105. This handbook is a good manual for conducting the systems engineering process. It is also cited as a good source for risk analysis, pp. 7, 37–44, June 1995.

National Transportation Safety Board (NTSB). Safety recommendation. National Transportation Safety Board 1990 [cited 14 December 2009]. Available from http://www.ntsb.gov/recs/letters/1990/A90_167_175.pdf, 2009.

Next-Generation SST: Technology Requirements, *Aerospace Engineering*, April 1994, pp. 29–31.

Oehmen, Josef, ed. *The Guide for Lean Enablers for Managing Engineering Programs*: Joint MIT-PMI-INCOSE Community of Practice on Lean in Program Management, 2012.

Oxford English Dictionary (OED). In *The Shorter Oxford English Dictionary on Historical Principles*, edited by C. T. Onions. Oxford: Oxford Univeristy Press. Original edition, 1933, 1973.

Page, Scott E. *Diversity and Complexity*. Princeton, NJ: Princeton University Press, 2011.

Pariès, Jean. Lessons from the Hudson. In *Resilience Engineering in Practice: A Guidebook*, edited by E. Hollnagel, J. Pariès, D. D. Woods and J. Wreathhall. Farnham, UK: Ashgate Publishing Limited, 2011.

Paté-Cornell, M. Elisabeth. Organizational Aspects of Engineering System Safety: The Case of Offshore Platforms, *Science*, Vol. 250, November 30, 1990, pp. 1210–1216.

Pirsig, Robert. *Zen and the Art of Motorcycle Maintenance*. New York: Bantam Books, 1974.

Perrow, Charles. *Normal Accidents: Living With High Risk Technologies*. Princeton, NJ: Princeton University Press, 1999.

Petersen, Thomas J. and Sutcliffe, Peter L. Systems Engineering as Applied to the Boeing 777, AIAA 1992 Aerospace Design Conference, Irvine, California, 1992.

Pyster, Arthur, ed. *Systems Engineering Body of Knowledge*. First edition. Stephens Institute, Hoboken, NJ and the Naval Postgraduate School, Monterery, CA, 2012.

RAF. Proceedings of a Board of Inquiry into an Aircraft Accident. Royal Air Force, 2007.

Ramo, Simon, 1973, quoted in Rechtin, 1991, p. 28.

Reason, James. *Human Error*. Cambridge, UK: Cambridge University Press, 1990.

Reason, James. *Managing the Risks of Organisational Accidents*. Aldershot, UK: Ashgate Publishing Limited, 1997.

Rechtin, Eberhardt. *Systems Architecting: Creating and Building Complex Systems*. Englewood Cliffs, NJ; Prentice-Hall. This book, written by Professor Rechtin of the University of Southern California, describes the process of synthesizing any system at the highest level of the system architecture, 1991.

Rijpma, Jos A. Complexity, Tight Coupling and Reliability: Connecting Normal Accidents Theory and High Reliability Theory, *Journal of Contingencies and Crisis Management* Vol. 5, No. 1, pp. 15–23, 1997.

Satchell, Paul. *Cockpit Monitoring and Alerting Systems*. Aldershot, UK: Ashgate Publishing Limited, 1993, pp. 10, 17, 52, 53, 59, 120.

Shafer, John B. Practical Workload Assessment in the Development Process, *Proceedings of the Human Factors Society 31st Annual Meeting*, Santa Monica, California, pp. 1408–1410.

Skybrary. Non Avian Wildlife Hazards to Aircraft. Eurocontrol, 16 October 2013 [cited 4 January 2014]. Available from http://www.skybrary.aero/index.php/Non_ Avian_Wildlife_Hazards_to_Aircraft?utm_source=SKYbrary&utm_campaign

=ee75eb4949-SKYbrary_Highlight_01_01_2014&utm_medium=email&utm_term=0_e405169b04-ee75eb4949-276526209, 2013.

Skybrary. 4D Trajectory Concept. Eurocontrol [cited 9 January 2014]. Available from http://www.skybrary.aero/index.php/4D_Trajectory_Concept? utm_source=SKYbrary&utm_campaign=b60834be85-SKYbrary_Highlight_06_01_2014&utm_medium=email&utm_term=0_e405169b04-b60834be85-276526209#Benefits_of_4D_Trajectory_Operations, 2014.

Skybrary. SE 120: Terrain Awareness and Warning System. Eurocontrol, 21 August [cited 4 January 2014]. Available from http://www.skybrary.aero/index.php/SE120:_Terrain_Awareness_and_Warning_System_(TAWS)_Improved_Functionality, 2014.

Sillitto, Hillary G. Design Principles for Ultra-Large-Scale Systems. In *International Council on Systems Engineering International Symposium*. Chicago, IL, 2010.

Society of Automotive Engineers (SAE). ARP 4754, Certification Considerations for Highly-Integrated or Complex Aircraft Systems Society of Automotive Engineers, 1996.

Society of Automotive Engineers (SAE). ARP 4754A, Guidelines for the Development of Civil Aircraft and Systems, edited by John Dalton: Society of Automotive Engineers, 2010.

Stevens, Richard, Peter Brook, Ken Jackson and Eliot Arnold. *Systems Engineering: Coping With Complexity*. London: Prentice Hall, 1998.

Software Considerations in Airborne Systems and Equipment Certification. RTCA/DO-178B, RTCA, Inc., 1 December 1992.

Specification for Manufacturers' Technical Data. Air Transport Association of America (ATA) Specification 100, Revision 28, 15 March 1989.

Specification Practices. Department of Defense, MIL-STD-490A, 4 June 1985. (superseded by MIL-STD-961D).

Specification Practices. MIL-STD-490A, October 30, 1968.

System Engineering, Department of the Army, US Army Field Manual, FM 770-78, April 27, 1979.

System Engineering. Electronic Industries Association (EIA), SYSB-1, December 1989.

Systems Engineering. Draft military standard, MIL-STD-499B, 6 May 1992 (cancelled).

Vaughn, Diane. *The Challenger Launch Decision: Risky Technology, Culture, and Deviance at NASA*. Chicago, IL: University of Chicago Press. Original edition, 1996, 1997.

Woods, David D. Essential Characteristics of Resilience. In *Resilience Engineering: Concepts and Precepts*, edited by E. Hollnagel, Woods, David D., and Leveson, Nancy. Aldershot, UK: Ashgate Publishing Limited, 2006.

Zarboutis, Nikos, and Peter Wright. Using complexity theories to reveal emerged patterns that erode the resilience of complex systems. Paper read at Second Symposium on Resilience Engineering, 8–10 November, at Juan-les-Pins, France, 2006.

Zimmermann, Kyla, Jean Pariès, René Amalberti, and Daniel H. Hummerdal. Is the Aviation Industry Ready for Resilience? Mapping Human Factors Assumptions across the Aviation Sector. In *Resilience Engineering in Practice: A Guidebook*, edited by E. Hollnagel, J. Pariès, D. D. Woods and J. Wreathhall. Farnham, UK: Ashgate Publishing Limited, 2011.

Glossary

Absorption	The capability of withstanding a design-level disruption
Airworthiness	The condition of an item (aircraft, aircraft system, or part) in which that item operates in a safe manner to accomplish its intended function (ARP 4754A, 2010)
Allocation [def. 1]	The assignment of a performance requirement to a function
Allocation [def. 2]	The assignment of a requirement to a system element
Allocation [def. 3]	The breakdown of a top-level requirement into its subordinate components, for example, weight
Analysis	A type of verification. Any kind of mathematical, computational, or logical task performed to verify a requirement which cannot be verified in any other manner. Includes in-service evaluation and similarity analyses
Anthropocentric	Pertaining to the human view, for example, the human view of a system
Anthropometry	An applied branch of anthropology, concerned with the measurement of the physical features of people (Chapinis, 1996)
Architecting	The process of determining the arrangement of the parts of a system
Architecture	The arrangement of the parts of a system or subsystem; this can apply either to the physical architecture or the functional architecture
Arousal	A form of stress where activation is resolved following termination of the stressor or perturbing event (Satchell, 1993)
Assurance	The planned and systematic actions necessary to provide adequate confidence that a product or process satisfies given requirements (DO-178B)

Blended wing-body (BWB) An aircraft configuration in which the wing and body are integrated into a single unit

Canard A horizontal surface located on the forward portion of the fuselage for improved control. Canards have rarely been used on modern aircraft, but were part of the Wright brothers' *Kitty Hawk*

Certification The legal recognition that a product, service, organization or person complies with the applicable requirements. Such certification comprises the activity of technically checking the product, service, organization or person, and the formal recognition of compliance with the applicable requirements by issue of a certificate, license, approval or other document as required by national laws and procedures (ARP 4754A, 2010)

Certification basis The set of particular standards and FARs on which the certification of an aircraft is based

Certification maintenance requirement (CMR) A required periodic task established during the design certification of the aircraft as an operating limitation on it

Change-based aircraft An aircraft for a specific customer which may have a large number of requested changes

Cluster analysis An analysis by which different parts of a system are shown to have functional affinity and hence constitute a potential subsystem; this technique has been shown to be useful in the architecting of a system

Cockpit resource management (CRM) Class of programs designed to reduce the number of incidents and accidents which are behavioral in origin. Focus on such activities as training and pilot selection processes

Common cause Event which bypasses or invalidates redundancy or independence; that is, an event which causes the simultaneous loss of several redundant or independent items

Common cause analysis Generic term encompassing zonal safety analysis, particular risk analysis, and common mode analysis. (ARP 4654A, 2010)

Complexity The state of a system characterized by many components, many interfaces, and variability in the relationship among the components; this term is also used to describing the difficulty in understanding such a system

Component	Any self-contained part, combination of parts, subassemblies, or units, which perform a distinctive function necessary to the operation of the system
Configuration index	A catalogue of the physical elements which comprise the aircraft and its subsystems
Configuration management	A process for controlling both the configuration of the aircraft and the data required to define the aircraft
Constraint	Any non-performance requirement. Constraints include weight, dimensions, environments, and any other factor which may constrain the design
Control	Technical SE management activities which occur during all program phases and at all levels of the aircraft hierarchy, such as configuration and risk management
Demonstration	A method of verification, similar to test except does not require sophisticated instrumentation
Derivative aircraft	An aircraft which utilizes major components of existing aircraft as the basis for the development of an aircraft which meets new requirements
Derived requirements	Requirements that are dependent on the design solution
Design	The result of the design process, as distinct from the requirements process
Design requirement	The design characteristic which is the product of the synthesis process
Development fixture	A mock-up of the aircraft used during development to assure that the spatial allocation for all components is correct and that they fit correctly
Disruption	Damage or loss of functionality resulting from an encounter with a threat
Drift correction	The capability of anticipating or detecting a disruption in advance and performing a corrective action
Electromagnetic interference	The disruptive interference caused by a magnetic field emitted by an electrical component, such as a generator
Electronic development fixture	An electronic, that is, computer-generated, version of a development fixture

Element
A generic term to describe any subdivision of the aircraft hierarchy. An element may be a segment, system, subsystem, or component. The aircraft itself is an element

Empennage
The entire tail assembly of an aircraft. May or may not include the tail cone depending on the practice of the manufacturer

Enterprise Systems Engineering (ESE)
The systems engineering of an entire enterprise, such as a commercial aircraft enterprise

Environment
The natural and induced conditions experienced by a system including its people, product, and processes (ANSI/EIA 632, 1999)

Examination
A type of verification. A visual confirmation that a requirement has been met. Also called inspection

Failure condition
A condition having an effect on the aircraft and/or its occupants, either direct or consequential, which is caused or contributed to by one or more failures or errors, considering flight phase and relevant adverseoperational or environmental conditions or external events (ARP 4754A, 2010)

Firmware
Any electronic device which contains imbedded programming logic

Flow-down
The process of passing or allocating any parameter, requirement, or function from a higher level of the aircraft hierarchy to a lower level

Function
A task, action, or activity performed to achieve a desired outcome

Functional allocation
Assignment of requirements to lower-level functions

Functional analysis
Examination of a defined function to identify all the subfunctions necessary to the accomplishment of that function

Functional hazard assessment (FHA)
A systematic, comprehensive examination of functions to identify and classify Failure Conditions of those functions according to their severity (ARP 4754A, 2010)

Fuselage
The body of an aircraft. May or may not include the nose or the tail cone depending on the practice of the manufacturer

Hierarchy	The layered arrangement of a system, especially the abstract hierarchy, or mental model of a system
Holistic	Treating the system as a whole taking into account the interactions among the elements as opposed to treating the elements separately
Human in the loop	The capability of having humans in the system where needed
IDEF0	Integrated Definition for Function Model, a method of functional analysis (see Reductionism)
INCOSE	International Council on Systems Engineering
Inspection	Alternative term for examination
Integrated product development (IPD)	A systematic management approach to the development of products such as aircraft and aircraft subsystems. Key elements of IPD include cross-functional integrated product teams (IPTs) and integrated and concurrent activities to develop products and processes
Interface	A boundary between two system elements.
Interface control drawing (ICD)	A document which captures all basic information about interfaces between two elements, including the type of interface (electrical, pneumatic, hydraulic, and so on) and the interface characteristics (functional or physical)
Inter-node interaction	The capability of two or more components of a system to interact with each other, such as communicate
Issue	A risk that has already been realized or whose consequence is inevitable regardless of the mitigation step
Item	One or more hardware and/or software elements treated as a unit (see Product)
Large-scale system integration	The integration of many systems containing complex interfaces
Latent fault or error	Design flaw that lies undetected until a catastrophic event occurs

Layered defense	The capability of having multiple means of withstanding a disruption; also called defense in depth
Limit degradation	The capability of arresting the degradation of the absorption capability due to aging or lack of maintenance
Margin	An capability of withstanding increased levels of a disruption due to uncertainty in the threat level
Master minimum equipment list (MMEL)	A document established by the manufacturer which lists what aircraft equipment can be inoperative (and under what conditions) and still fly the aircraft safely (see MEL)
Mean time between failures (MTBF)	Mathematical expectation of the time interval between two consecutive failures of a hardware item. NOTE: The definition of this statistic has meaning only for repairable items. For non-repairable items, the term Mean Time To Failure (MTTF) is used (ARP 4754A, 2010)
Mean time between unscheduled removals (MTBUR)	Time interval between two consecutive unscheduled removals of an item. An unscheduled removal is a removal of an item brought about as a result of a known or suspected malfunction and/or defect
Metric	A measure, usually quantitative, of the value of a process, such as SE
Minimum equipment list (MEL)	A document established by the airline which lists what aircraft equipment can be inoperative (and under what conditions) and still fly the aircraft safely; a subset of the MMEL (see MMEL)
Needs	Those desires usually from the commercial aircraft customer that will result in an economically viable aircraft. These needs will be translated into verifiable product requirements
Off-the-shelf	Pertaining to a commercially available product which meets the specified requirements
Performance requirement	The extent to which a mission, operation, or function must be executed, generally measured in terms of quantity, quality, coverage, timeliness, or readiness (ANSI/EIA 632, 1999)
Peripheralization	A complex psychological state which results from a shift in the pilot role from direct contact and control of the aircraft to one of system monitor (Satchell, 1993)

Preliminary system safety assessment (PSSA)	A systematic evaluation of a proposed system architecture and its implementation, based on the Functional Hazard Assessment and Failure Condition classification, to determine safety requirements for systems and items (ARP 4754A, 2010)
Process assurance	The set of activities that ensure that the development of the aircraft, its subsystems, and the supporting processes are appropriate, maintained, and followed
Product	A generic term used for any item, either hardware or software, which will be the end result of the SE process. A product can be a system, subsystem, or component. The term "item" is also used
Product Systems Engineering (PSE)	The systems engineering of products containing hardware or software, such as an aircraft
Reductionism	The concept that the parts of a system should be treated separately so that a system is built up of the sum of its parts (see Holism)
Redundancy, functional	Having two independent and physically different means to accomplish a function, also called design diversity
Redundancy, physical	Having two independent and physically identical means to accomplish a function, also called design redundancy
Reliability	The probability that an item will perform a required function under specified conditions, without failure, for a specified period of time (ARP 4754A, 2010)
Reorganization	The capability of restructuring a system when needed to recover from a disruption
Requirement	A statement of required performance or design constraint to which a product must conform. A requirement must be verifiable
Resilience	The capability of anticipating or detecting a disruption, surviving that disruption, and recovering all or part of the initial functionality
Risk	An undesirable situation or circumstance that has a realistic probability of occurring and an unfavorable consequence

Safety   The state in which risk is acceptable (ARP 4754A, 2010)

Segment   A major collection of aircraft equipment. A segment may be the wing, a subsystem, or simply a collection of elements having similar functions

Service Systems Engineering (SSE)   The systems engineering of a service, such as the maintenance of an aircraft

Similarity   In systems engineering a type of verification by analysis. In ARP 4754A (2010) applicable to systems similar in characteristics and usage to systems used on previously certificated aircraft. In principle, there are no parts of the subject system more at risk (due to environment or installation) and that operational stresses are no more severe than on the previously certificated aircraft

Software   Computer programs, procedures, rules, and any associated documentation pertaining to the operation of a computer system (ARP 4754A, 2010)

Specialty requirement   A requirement set by one of the various specialty engineering disciplines, such as human factors, reliability, maintainability, safety, environments, mass properties, and software. Specialty requirements can either be performance requirements or constraints

Specification   The collection of requirements which, when taken together, constitute the set of criteria which define the functions and attributes of an item (ARP 4754A, 2010)

Stress   The emotional state, either detrimental or beneficial, which results from various stressors

Stressor   Any stimulus which may result in stress

Subsystem   A subdivision of the aircraft hierarchy of Figure 2.1, one level below the aircraft. The elements of a subsystem, when viewed together, satisfy the definition of a system. Traditionally called a system in the aircraft industry

Subsystem-level   Pertaining to functions, requirements, or trade-offs *within* a given subsystem

Supply chain   The collection of suppliers of commercial aircraft suppliers and their relationships

Synthesis	The translation of input requirements into possible solutions satisfying those inputs (ANSI/EIA 632, 1999)
System	An interacting combination of elements, viewed in relation to function (official INCOSE definition). In ARP 4754A (2010) and DO-178B (1992), for example, refers to a subsystem
System analysis	Trade-offs and other activities leading to a system synthesis
System safety assessment (SSA)	A systematic, comprehensive evaluation of the implemented system functions to show that relevant safety requirements are met (ARP 4754A, 2010)
System synthesis	The process of creating a design. System synthesis begins with the development of the system architecture (Section 2.3) and the system functions (Section 3.2) and ends with the assignment of hardware and software to the requirements. System synthesis is discussed in Chapter 7
Systems architecting	A process for creating unprecedented, complex systems. Focuses on six core concepts or ideas: the systems approach, purpose orientation, ultraquality, modeling, experienced-based heuristics, and certification (Rechtin, 1991)
Systems engineering	The interdisciplinary approach and means to enable the successful realization of successful systems (official INCOSE definition) (see System)
Systems Engineering Body of Knowledge (SEBoK)	An electronic compendium of facts about systems engineering, in Wiki format
System validation	The assurance that the entire system meets its mission objectives
Technical performance measurement (TPM)	The continuing verification of the degree of anticipated and actual achievement of technical parameters (ANSI/EIA 632, 1999)
Test	A type of verification which requires instrumentation. Includes both laboratory and flight tests
Top-level	Pertaining to the highest level of either the aircraft or aircraft system as defined by the aircraft system architecture of Figure 2.1. Also pertains to analyses or relations among two or more subsystems

Traceability   The recorded relationship established between two or more elements of the development process, for example between a requirement and its source or between a verification method and its requirement (ARP 4654A, 2010). In systems engineering traceability more commonly refers to the characteristic by which requirements at one level of a design may be related to requirements at another level. Traceability also encompasses the relationship between a performance requirement and the function from which the performance requirement was derived

Trade-off   An analysis conducted to determine the preferred option among two or more options, such as the number of engines, based on a figure of merit, such as cost or weight. Trade-offs can be either top-level or subsystem-level

Turnaround time   The time between arrival at a gate to departure from a gate of an aircraft

Validation   The determination that the requirements for a product are sufficiently correct and complete (ARP 4754A, 2010) (see Product) (see also System validation)

Variability   The lack of stability in the relationship between the components of a system; a contributor to the complexity of the system; also known as information entropy

Verification   The evaluation of an implementation of requirements to determine that they have been met (ARP 4754A, 2010)

Vigilance   Sustained attention, or the ability of observers to maintain their attention to remain alert to stimuli over prolonged periods of time (Satchell, 1993)

Winglet   A vertical surface located on the wingtip of an aircraft to reduce the wingtip vortex effect and thus improve lift. Winglets have become common on modern aircraft

Workload   The number of things an operator has to do within any particular time period modified by their level of difficulty (Shafer, 1987)

Index